Birth of the Shinkansen

Tetsuo Shimomae

Birth of the Shinkansen

The Origin Story of the World-First
Bullet Train

Tetsuo Shimomae
Electrical Engineering
Japan Railway Electrical Engineering
Association
Tokyo, Japan

ISBN 978-981-16-6540-0   ISBN 978-981-16-6538-7   (eBook)
https://doi.org/10.1007/978-981-16-6538-7

Cover illustration: © Anesthesia/PIXTA

This Springer imprint is published by the registered company Springer Nature Singapore Pte Ltd.
The registered company address is: 152 Beach Road, #21-01/04 Gateway East, Singapore 189721,
Singapore

Photo 1   Rolling test rig for 1/10 scale model (1957) (provided by RTRI)

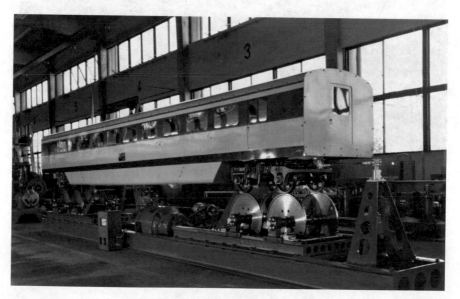

Photo 2   Rolling test rig for 1/5 scale model (1959) (provided by RTRI)

Photo 3 1/10 scale models for wind-tunnel experiment (1954) (provided by RTRI)

Photo 4 Wind-tunnel test of 1/12 scale Shinkansen lead car model (1962) (provided by RTRI)

Photo 5 Brake test machine (provided by RTRI)

Photo 6 Track bed test machine (provided by RTRI)

Photo 7　Face of the test car (1962) (provided by RTRI)

Photo 8　Running test (1963) (provided by RTRI)

Photo 9   Measurement work in a test train (1963) (provided by RTRI)

Photo 10   Measurement work at the trackside (1963) (provided by RTRI)

Photo 11 Series 0 twelve-car train about to open for business, Tokyo station (1964) (provided by RTRI)

Preface

On October 1, 1964, the world's first high-speed commercial train, known as the Shinkansen (which means "new trunk line"), began operating. It was the first in the world to operate at a speed of more than 200 km/h. In Europe at the time, the maximum operating speed for trains was 160 km/h, and the general consensus was that maximum speeds of 200 km/h were unrealistic. Therefore, the start of the Shinkansen operation was an epoch-making event that opened up a new world of railways.

The Shinkansen came into being when railways were deemed to be an unstable industry worldwide. In Japan, many people believed that the country should invest in highways and aircraft, which were thought to be the travel modes of the future, rather than the outdated railways. However, the success of the Shinkansen changed public opinion, and since its opening, the Shinkansen has continued to grow; as of 2016, the Shinkansen line was 3,000 km long.

Inspired by the success of the Shinkansen, European countries began to develop their own high-speed trains. In the UK, in 1976—twelve years after the opening of the Shinkansen—the IC125 train began operating at a maximum speed of 201 km/h . The TGV began operating in France in 1981 (260 km/h), the ETR in Italy in 1988 (200 km/h), the ICE in Germany in 1991 (280 km/h), and the AVE in Spain in 1992 (300 km/h). In Asia, technology transfer from Europe and Japan led to the opening in 2004 of the KTX in Korea, and in 2007, the CRH in China and the 700T in Taiwan. More than 20 countries now operate high-speed trains.

This text describes how the Shinkansen was realized through technological developments and how it has developed up to the present day. The technological history of the Shinkansen can be divided into four stages:

Stage 1: Thirteen years from 1945, when World War II ended, to 1958, when the decision was made to build the Tokaido Shinkansen
Stage 2: Six years between the decision to build the Shinkansen and its opening in 1964
Stage 3: Twenty-three years from the opening of the Shinkansen to the breakup of the Japan National Railways into nine companies and one incorporated foundation in 1987
Stage 4: Thirty-three years from the breakup of JNR to the present

The Shinkansen was born in stages 1 and 2 and grew up in stages 3 and 4.

As outlined in Part II of this text, today's Shinkansen trains have evolved beyond comparison with the first in speed, safety, ride quality, train frequency, on-time performance, and environmental friendliness. The most significant technological developments occurred in the years leading up to Shinkansen's opening, and therefore, stages 1 and 2 are discussed in extreme detail.

This text consists of three parts:

Part I contains five chapters that describe stages 1 and 2.

- Chapter 1 provides the basic research on railway technology conducted mainly in the absence of computers by engineers who moved from the military to the National Railways after the war.
- Chapter 2 presents the lecture held in May 1957 to commemorate the 50th anniversary of the Railway Technical Research Institute (RTRI), which played a significant role in promoting the Shinkansen project.
- Chapter 3 describes the process leading to the decision to build the Shinkansen.
- Chapter 4 discusses the extensive technical developments necessary in realizing the Shinkansen, which was carried out on a very tight schedule to ensure that the Shinkansen would be ready for the 1964 opening of the Tokyo Olympics.
- Chapter 5 briefly describes the Shinkansen operations in the first few years after its opening.

Part II consists of Chap. 6 that overviews stages 3 and 4, an introduction of today's highly evolved Shinkansen, including high frequency of operation, maximum speed, high on-time performance, high level of safety, effective measures against big earthquakes, and high environmental friendliness. Chapter 6 also provides an overview of high-speed rail accidents worldwide and delineates factors that have protected the Shinkansen from such disasters.

Part III showcases photographs of the Shinkansen trains from 1964 to 2020.

Tokyo, Japan   Tetsuo Shimomae
February 2021

Acknowledgments

This text would not have been possible without the help of many people.

I would like to express my gratitude to the Railway Technical Research Institute (RTRI), Central Japan Railway Company (JR Central), NIPPON SHARYO, LTD., SNCF, La Vie du Rail, The Center of the Tokyo Raids and War Damage, Yokosuka City Chuo Library, Japan Transport Safety Board of Ministry of Land, Infrastructure, Transport and Tourism (MLIT), The Japan Society of Mechanical Engineers, Japan Railway Electrical Engineering Association, Mr. Shinichi Tanaka, Mr. Saburo Kiyosawa, Mr. Tatsuya Miyasaka, and photographers in the Creative Commons and stock photo Web sites for providing the pictures for this text. Most of the historic photos in Part I were supplied by the RTRI. JR Central also enriched Part III with many beautiful pictures.

I am also grateful to Mr. Murray Hughes for the permission to quote from his book.[1]

Part I of this book is a reconstructed version of a book, *Technology Developments in Realizing the Shinkansen*, that I published in Japanese in 2019 via Seizando Publishing. Approximately half of that original publication appears in this text and has been revised to accommodate readers worldwide. I would like to thank Seizando Publishing for understanding my wish to share with people worldwide the history of technological developments that made high-speed railways possible.

Special thanks are due to Ms. Swati Meherishi, Editorial Director at Springer Nature, for taking up my book proposal. She gave me a lot of advice and guided me through the publication process as I was not used to publishing abroad. I would also like to thank Ms. Muskan Jaiswal, Assistant Editor at Springer Nature, who carefully analyzed the manuscript and helped me with processes and paperwork.

Writing a book in English was a daunting task for me as I do not use English in my daily life. I am indebted to Ms. Rose Kernan and Ms. Patricia Daly of RPK Editorial Services, who checked the manuscript and spent a lot of time and effort to bring it to a publishable level.

---

[1] Murray Hughes, *The Second Age of Rail-A History of High-Speed Trains,* 2nd Edition, The History Press, 2020.

Finally, I am thankful that I could have been keeping my physical and mental health until an advanced age and publish this book. I owe this to the DNA that my parents left me and my wife Yukiko, who has been taking care of my health for many years, for which I shall be ever grateful.

July 2021   Tetsuo Shimomae

Contents

About the Author

Tetsuo Shimomae received his Master's degree in Electrical Engineering from Nagoya Institute of Technology in 1966 and joined the Japanese National Railways (JNR) the same year. In 1967, he started working with the Railway Technical Research Institute (RTRI), developing an overhead catenary equipment inspection vehicle and improving the Shinkansen pantograph. In 1974, he received the Ohm Award from the Promotion Foundation for Electrical Science and Engineering. In 1987, when JNR was split into several private companies, he belonged to the JR Central Company which took over the Tokaido Shinkansen. There, he served as head of the Shinkansen division's electrical department, deputy general manager of the Engineering Department, manager of the Safety Management Department (director), and Shizuoka Branch President (managing director). After leaving the company, he served as president and chairman of Shinsei Technos Co., Ltd., a subsidiary of JR Central in charge of construction and maintenance of railway electrical equipment. In 2008, he became chairman of the Japan Railway Electrical Engineering Association whose members are railway electrical engineers, manufacturers, and associated construction companies, and has been an advisor since 2012.

Part I

The Birth of the Shinkansen

On August 15, 1945, World War II came to an end with the defeat of Japan. At the end of October, the Imperial Japanese Navy's Aviation Technology Center, where Matsudaira[1] had worked for 11 years, was disbanded.

In search of a new position, he visited Ikeda[2] in November of that year, who was the manager of the railcar division of the Railway Technical Research Institute (hereafter referred to as RTRI). Ikeda listened to his story and arranged for him to work at the RTRI.

Thus, at the age of 35, Matsudaira moved to the RTRI, where he would later make notable contributions to the realization of high-speed railways. What he did will be discussed later, but first, let us look at who he was.

In March 1934, after graduating from the Department of Shipbuilding Engineering at Tokyo Imperial University (now University of Tokyo), Matsudaira entered the Naval Aviation Technology Center.

At the time, airplane vibrations were major problems in the Navy, but since there were no experts in that field, he was to study vibrations. However, no systematic lectures on mechanical vibrations were given in any Japanese university, so he began studying mechanical vibrations on his own. The main source of his study was the textbook *Mechanical Vibrations* by Jacob Pieter Den Hartog, a professor at MIT at the time.

In his memoirs, Matsudaira wrote that he was completely fascinated by vibrations, and when he finished the *Mechanical Vibrations* text he felt that he was, thanks to this masterpiece, already a full-fledged expert on it. He then obtained a

---

[1] Tadashi Matsudaira joined the Imperial Japanese Navy in 1934 and moved to RTRI in 1945, first as a senior researcher; he later became the head of the vehicle dynamics laboratory of RTRI and finally director of RTRI.

[2] Shojiro Ikeda; later a member of the Science Council of Japan.

Photo 1 Tokyo in disrepair, 1945 (provided by The Center of the Tokyo Raids and War Damage)

Photo 2 Japanese Navy Aviation Technology Center, 1939. (provided by Yokosuka City Chuo Library)

classic British literature on flutter[3] and began to study this phenomenon. The British document revealed that the frequent airplane breakdowns during World War I were caused by self-excited vibrations called fluttering and described fluttering's fundamental nature and how to prevent it.

Based on these findings, in 1936, Matsudaira implemented countermeasures against fluttering on Type 92 and Type 95 land attack aircraft. These accidents were not serious because of the low speeds of 150–200 knots (approximately 280–370 km/h), but a Zero fighter plane accident in 1941 at about 320 knots resulted in disintegration in midair and the pilot's death. More than 150 Zeroes were in production at the time, so the accident was a significant shock to the Navy.

---

[3] Frazer, R. A., & W. J. Duncan, "The Flutter of Aeroplane Wings," *R.&M.*, no. 1155, 1928.

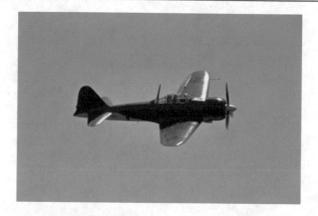

Photo 3 Zero fighter (provided by Kumachan/PIXTA)

At the time, no country in the world had the wind-tunnel testing technology to estimate the speed at which flutter begins, so Matsudaira's team began to investigate the flutter phenomenon at high speeds by building a 1/10 scale wing model and using a wind tunnel. The bending stiffness and mass distributions of the model were designed to be mechanically similar to the real thing.

In his memoirs, Matsudaira described the wind-tunnel test as follows: [1]

> As the wind speed increased, the auxiliary wing suddenly began to vibrate in small increments. Looking closely, I could see that the main wings are twisting and vibrating. It was precisely a wing-twisting flutter. This vibration, as I expected, was just what caused the accidents so far! I felt a sense of relief that the cause of the accident had been found, but at the same time, I felt a sense of regret for my lacking ability and a sense of serious responsibility as the person in charge of vibration testing of aircrafts. I still cannot forget the complicated and painful feelings I had at that time.
>
> Thus, the accident was wholly resolved within two months after it occurred. The solution was to increase the mass of the auxiliary wing to balance against the torsional vibrations of the main wing dynamically. (Author's translation of the Japanese)

As described, Matsudaira was an engineer who solved the vibration problem, which cannot be tested in real size, by analytical methods and model experiments.

Basic Study of Railways After World War II

1

1.1 Vibration and Hunting Analysis of Railcars

In December 1946, exactly one year after Matsudaira moved to RTRI, Hideo Shima,[1] then manager of the Ministry of Transportation's power railcar division, started a study group to improve electric railcars' ride quality. The participants were the Ministry of Transportation's technical staff, including Shima, RTRI's engineers, including Matsudaira, and the vehicle manufacturers (Photo 1.1).

In his memoirs, Matsudaira wrote [2]:

> At that time, the vibrations of railcars were several times greater than those of today, and those of electric railcars, in particular, were so great that the ride was extremely uncomfortable. Therefore, it was unthinkable to use such electric railcars for long-distance trains. However, Shima was an advocate of electric railcar trains at that time. He thought that he should improve the electric railcars' vibrations thoroughly to realize his idea, so he asked the author to reduce their vibrations. (Author's translation of the Japanese)

The study group met six times, the last time in April 1949 (Table 1.1).

The purpose of the study group was to improve the ride quality of conventional electric railcars, but the basic study of the railcar's vibration characteristics led directly to the realization of the later Shinkansen vehicle. Table 1.1 shows the activities of the group (the mark O indicates one paper submitted).

---

[1] Hideo Shima joined the Ministry of Transport in 1925. He later became director of the vehicle department, chief engineer of Japanese National Railways, and finally president of National Space Development Agency of Japan.

© The Author(s), under exclusive license to Springer Nature Singapore Pte Ltd. 2022
T. Shimomae, *Birth of the Shinkansen*,
https://doi.org/10.1007/978-981-16-6538-7_1

Photo 1.1 Matsudaira
(when he was the director of
RTRI) (provided by RTRI)

1.1.1 The First Meeting

a. **Natural Frequency Analysis of Railway Vehicles**

Matsudaira presented a paper on the analysis of rail car's natural frequencies at the
first meeting. He wrote in the paper [3]:

> It is of fundamental importance to clarify the natural frequencies of railway vehicles in
> reducing the vibration. For a swing hanger type bogie [see Fig. 1.1], Dr. Muto and Dr.
> Nakajima have already shown how to calculate the railcar's natural frequencies . However,
> their formulas are theoretically uncertain partially and too complex to be used by designers.
> Therefore, the natural frequencies have rarely been considered in the design of railcars in
> the past. (Author's translation of the Japanese)

Figure 1.1 shows the structure of a swing hanger-type bogie, which was com-
monly used at the time. In this bogie, the axle supports the bogie frame via the
primary suspension, and the bogie frame hangs the lower bolster by the swing
hangers so that the lower bolster can move from side to side. The lower bolster
supports the upper bolster via the secondary suspension, and the upper bolster
supports the car body with the center pivot and side bearers. Therefore, the weight
of the car body is transmitted to the rails as follows:

Car body → center pivot and side bearers → upper bolster → secondary suspension
→ lower bolster → swing hangers → bogie frame → primary suspension → wheel → rail.

The primary and secondary suspensions mitigate vertical vibrations, and the
swing hanger system handles lateral vibrations.

Photo 1.2 shows a MOHA63-type electric railcar with the swing hanger bogie
(MO means power cars and HA means second class cars).

Setting the coordinate axes as shown in Fig. 1.2, Matsudaira classified the
motion of railcars into six modes: back and forth, up and down, left and right,
rolling, pitching, and yawing. Among these motions, the lateral and rolling motions
do not exist independently but appear in conjunction.

Table 1.1 Activities of the study group

Subject / Meeting		1st Dec. 1946	2nd Jul. 1947	3rd Dec. 1947	4th Jun.1948	5th Nov. 1948	6th Apr. 1949
Vibration	Vibration analysys of car bodies	OOOO OOOO	OOOOO	OOOO	OOOOO	OOOO	OOOO OOO
	Vibration analysys of swing hanger type bogies	OO	OOO	OOOO			OOO
	Track deflection		O				
	Riding quality		U				O
	Vibration measurement		O	OOO	OOOOOO	OOOO OOOO	OOOO O
	Vibrators for testing			OO	OO		
	Characteristics of layered leaf springs		OO	OOO	OOOOOO		OO
	Characteristics of coil springs				O		
	Oil dampers			O	O	O	O
	Air dampers and air springs			O	O	OO	
Hunting	Hunting analysis			OO			
	Creep force		O	OO	O		
	Scale model experiments			O	O		OO
Motor suspensions					OOO	O	OOOOO
Foreign bogies			O		O		OO
New bogies				O	O		

He analyzed the natural frequencies of vertical motion, pitching, and yawing using the models shown in Figs. 1.3, 1.4, and 1.5.

The analysis of vertical, pitching, and yawing motions using these models is not difficult, but the analysis of lateral motion coupled with rolling is complex. He modeled the vehicle as shown in Fig. 1.6 to analyze this motion, where m is the body mass, i_1, i_2, i_x, i_o are the rotational radii around O_1, O_2, G, O, and G is the gravity center of the car body.

Setting three centers of rotation, O_1 (due to the primary suspension), O_2 (due to the secondary suspension), and O (due to the swing hanger system), he derived the equations for this complex motion. Since there had been no established solution for this motion, there was an active debate on how to analyze it.

Matsudaira calculated the natural frequencies of the MOHA63-type railcar using the aforementioned analysis (Table 1.2).

This was the first time in Japan that the natural frequencies of a railcar were calculated.

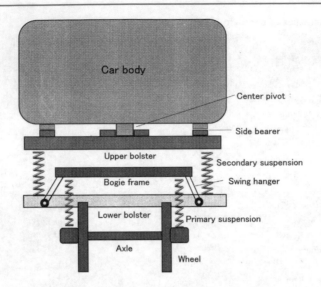

Fig. 1.1 Structure of swing hanger-type bogie

Photo 1.2 MOHA63-type electric railcar (Author)

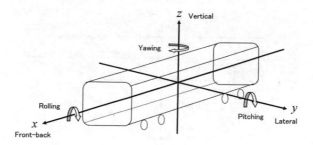

Fig. 1.2 Coordinate axis

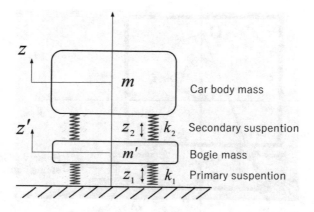

Fig. 1.3 Model for vertical motion analysis. *Source* Matsudaira, T., "Natural Frequencies of Coaches and Electric Railcars", RTRI Research Materials, no. 2, RTRI, 1949, p. 4

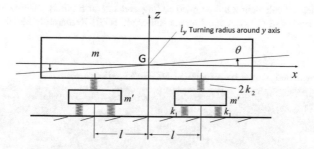

Fig. 1.4 Pitching calculation model. *Source* Matsudaira, T., "Natural Frequencies of Coaches and Electric Railcars", RTRI Research Materials, no. 2, RTRI, 1949, p. 5

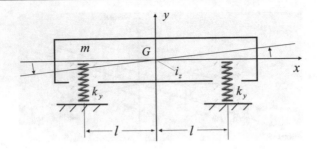

Fig. 1.5 Yawning calculation model. *Source* Matsudaira, T., "Natural Frequencies of Coaches and Electric Railcars", RTRI Research Materials, no. 2, RTRI, 1949, p. 10

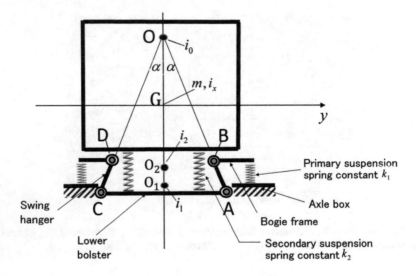

Fig. 1.6 Model for calculating the coupled rolling and lateral motions. *Source* Matsudaira, T., "Natural Frequencies of Coaches and Electric Railcars", RTRI Research Materials, no. 2, RTRI, 1949, p. 5

Table 1.2 Calculated natural frequencies of MOHA63 railcar

	Secondary suspension with no friction	Secondary suspension with very large friction
Vertical motion	104/min	136/min
Pitching	116	148
Rolling (Low order mode)	32	58
(High order mode)	67.5	73.5
Yawing	59	67

In order to verify these values, vibration tests were needed. However, this type of test had never been done before, and of course, there was no vibrator. In the end, Matsudaira and his coworkers ended up managing to make a vibrator.

As an interesting sidenote, there was a noteworthy description in a paper submitted by a vehicle manufacturer who attended this meeting:

> Those of us who are engaged in the manufacture of railway rolling stock always try to make reliable products, but it is not only important to make strong products. For freight cars, riding comfort may not be so important, but for passenger cars, designing a car that rides well is a major challenge for the future." (Author's translation of the Japanese)

In other words, at that time, railcars were made without consideration for riding comfort.

1.1.2 The Second Meeting

a. Vibration Test of MOHA63 Railcar

Matsudaira conducted a vibration test using a broken motorcycle engine as the vibrator, with unexpected results. The kind of vibration was the same as in the analysis, but the frequencies were utterly different. He attributed the difference to the following:

- The first mistake in the theoretical analysis was that the analysis did not take into account friction. In a real vehicle, there is friction in joints and sliding parts. The friction in layered leaf springs is particularly high.
- Second, the center pivot was assumed to be rigid, but in fact, it has spring action.
- Third, the secondary suspension was considered rigid laterally, but in fact, it can move laterally.
- Other possible causes are the low bending and torsional rigidity of the body.

In addition, the following information was revealed:

- The layered leaf spring is very hard dynamically compared to statically and does not function as secondary suspension.
- The bogie has more friction in sliding parts than necessary.

Due to many nonlinear elements, the natural frequencies cannot be calculated correctly, so meeting attendees considered the use of coil springs and dampers instead of layered leaf springs. The following exchange from the meeting transcripts reflects this development:

Photo 1.3 Layered leaf
spring used for secondary
suspension (Author)

Manufacturer: Can coil springs be used for secondary suspension if dampers are
used in combination with them?
Matsudaira: Yes.
Shima: I heard that in the U.S., most railcars use both coil springs and dampers.

b. **Beginning of Hunting Motion Study**

Traditionally, Japanese engineers understood hunting motion as the so-called
geometric hunting, derived by Klingel[2] in 1883, which states that the wavelength of
hunting motion is determined solely by the wheel diameter, wheel tread slope, and
rail spacing. However, the motion is not as simple as geometric hunting since the
wheels are subject to inertia, gravity, and creep force from the rails.

Realizing that the geometrical hunting theory was incomplete from the standpoint of vibrational dynamics, Matsudaira conducted a literature survey and found
that these forces had been taken into account in hunting analysis in Europe and the
USA and that they had been utterly untested in Japan.

At this second meeting, Matsudaira presented a paper titled "An Introduction to
the Hunting Motion Theory of Railcars, Incorporating the Creep Force." He
explained that creep force is the force generated at the contact surface of the wheel
and the rail by the elastic deformation of them, and that the hunting motion is an
unstable self-excited oscillation.

The following questions from the meeting transcripts indicate that Matsudaira's
research on hunting motion began around this time.

---

[2] Klingel, J., "Über den Lauf von Eisenbahnwagen auf gerader Bahn", *Organ fürdie Fortshritte
des Eisenbahnwesens*, no. 20, 1883.

Manufacturer: Are wheel diameter and wheel material factors in the creep force?
Matsudaira: The contact length and the radius are factors.
Shima: I think Friction force = Constant × Creep coefficient × Wheel load. This is a very interesting phenomenon, so I hope Matsudaira will develop the research on hunting.

The following conversation reveals the Japanese situation at the time, which was under the control of the General Headquarters of the Supreme Commander for the Allied Powers (GHQ):

Manufacturer: I can read a railway-related magazine in the library of the Occupation Forces. I want the Ministry of Transportation to talk to the Occupation authority so that I can read more of them.
Shima: At the moment, it's just *Railway Age*, but it looks like there will be much more available soon. I hope you can read the *Railway Gazette*.

At the end of the meeting, Shima said,

Vehicle's strength is just as important as the vibration, but we haven't started a workshop yet. I want Miki[3] to drive it.

At the time, it was not known what kind of force was applied to vehicle bodies. Knowing the force acting on the parts is essential not only for rolling stock but also for tracks, bridges, and overhead catenary equipment. This was made possible by Kazuo Nakamura.[4] His handmade wire strain gauges made it possible to measure wheels' dynamic loads and stresses in various parts of car bodies (see Sect. 1.2.1).

A participant's request for the place to hold the next meeting conveys the food shortage at that time: "Why don't you hold the next meeting in a place where there is plenty of food?".

1.1.3 The Third Meeting

a. Optimization of Suspension Constants

As mentioned, Matsudaira's formula for calculating the natural frequencies of railcars was not satisfactory. Therefore, he presented an analysis of when the secondary suspension has friction damping device (snubber) in parallel (Fig. 1.7).

---

[3] Tadanao Miki joined the Japanese Imperial Navy in 1933 and worked as an airplane designer at Naval Aviation Technology Center, moved to RTRI in 1945 as a senior researcher , and later became head of the railcar body laboratory of RTRI.
[4] Kazuo Nakamura joined the Imperial Japanese Army in 1942 and moved to RTRI from Nakajima Aircraft Co. Ltd. in 1946 as a senior researcher.

Fig. 1.7 Vertical vibration model of a bogie with friction damping device. *Source* Matsudaira, T., Kunieda, M. et al., "Theory of Forced Vibration of Bogie Cars with Special Reference to Optimum Values of Stiffness of Springs and Friction", RTRI Research Materials, no. 8, RTRI, 1951, p.4

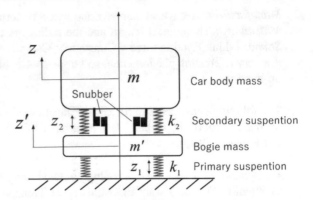

In that analysis, he states [4]:

> Recently, in the U.S., coil springs are mainly used in conjunction with snubbers (friction dampers) to provide damping to the secondary suspension. In order to minimize vertical vibrations of the car body, how should we select the stiffness of the primary and secondary suspension, and what level of friction should be applied to the secondary suspension? These questions are of fundamental importance in bogie design, but they have been determined empirically in the past and have no clear theoretical basis. Therefore, assuming that the secondary suspension has constant friction independent of its amplitude and frequency of vibration, I will clarify the vertical forced vibration characteristics of a railcar theoretically for the most basic case where the car runs on a sinusoidally bumpy rail, and determine the optimal values of spring stiffness and friction. (Author's translation of the Japanese)

From this analysis, Matsudaira showed that there exists a ratio between the constants of the primary and secondary suspension that minimizes the vertical vibration of the vehicle, and he made a diagram to find it. From the diagram, it was shown that the optimum suspension constants for a vehicle weighing 10 tons (empty) or 14 tons (full) with a bogie weighing 2.5 tons would be as follows:

- Primary suspension's constant (per side of bogie): 115 kgf/mm
- Secondary suspension's constant (as above): 72.4 kgf/mm
- Secondary suspension's friction force (as above): 450 kgf

Suspension optimization was achieved in 1952, after the study group was finished, with the DT17 bogie, which significantly improved the ride quality of the 80-series railcars described later.

b. **Hunting Motion Analysis**

In the second meeting, Matsudaira talked about the necessity of studying the dynamics of hunting motion, and in the third meeting, he submitted a paper on this topic. The paper's outline is introduced below; surprisingly, he was able to put it together in only five months, which indicates his outstanding ability in vibration analysis.

In the introduction to the document, Matsudaira states [5]:

The vibration that has the worst effect on the ride quality of railcars, especially of electric railcars, is the lateral vibration due to the hunting motion of the bogie. Therefore, in order to make comfortable railcars, the most important thing is to prevent or reduce the bogie's hunting behavior. However, little is known about the method for preventing or mitigating it. Therefore, this paper presents a theoretical analysis of hunting behavior of a 2-axle bogie to provide at least a part of the method to prevent it. The method of the analysis is based on creep theory to form linear equations of motion for the lateral and angular motion of the bogie and to examine the stability of the motion at small displacements. At first, the essence of the hunting behavior of a single wheelset is determined, then the hunting behavior of a 2-axle bogie with a rigid structure is considered. After that, the effects of the elastic support of the car body and the lateral movement of the axles to the bogie frame (elastic displacement or rattling), which are the most important practical issues, are examined. (Author's translation of the Japanese)

(1) Hunting Motion of a Single Wheelset

In Fig. 1.8, when the axle center G is at the middle of both rails as shown in (a), both wheels will travel the same distance as they rotate. However, if the axle moves to the left, as shown in (b), the left wheel advances more than the right because the left wheel's diameter is larger than the right. As a result, the axle moves at an angle, as shown in (c). When the axle continues to move further, the center of the axle returns to the middle of the rail, goes beyond the middle point, and moves to the right. As a result, the axle moves in a sinuous motion, as shown in Fig. 1.9.

In this case, if the wheel axis rotates slowly, the wavelength of the hunting motion, S_1, is approximated by Eq. 1.1.

$$S_1 = 2\pi \sqrt{\frac{a\gamma}{\lambda}}$$

(1.1)

where a, γ, λ are the symbols shown in Fig. 1.8.

Equation 1.1 was derived by Klinger, as described previously.

This equation is entirely correct when the wheelset is traveling very slowly. However, when the speed is high, the frequency of the oscillation of the hunting motion increases, resulting in the increase of the wheelset inertia in proportion to the square of the oscillation frequency, so this effect must be taken into account. In addition, the so-called creep force at the contact surface between the wheel and rail must be considered.

Since Matsudaira began his research on hunting motion shortly after the war, he could not refer to studies being conducted in Europe and the USA. He proceeded with his investigation from scratch and on his own.

The following is a description of how he solved the hunting problem, a key technology for high-speed railways, in an age when computers were not available.

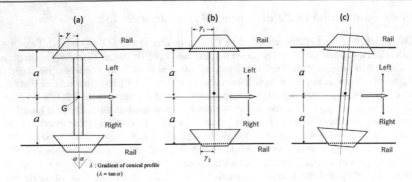

Fig. 1.8 Single wheelset on the rail

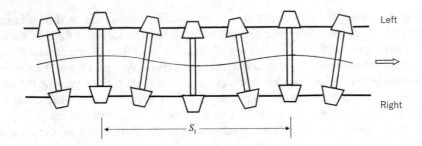

Fig. 1.9 Hunting motion of a single wheelset

(2) **Hunting Motion Analysis of Two-Axle Bogie**

When two wheelsets are connected at the center, as shown in Fig. 1.10, they are restrained from each other, so they cannot move freely, as in the case of a single wheelset. The general expectation is that the movement of two wheels will be less likely to develop into hunting than with one wheelset.

In this case, each wheel generates slippage as it rolls. Slippage caused by elastic deformation of the wheels and rails at the point of contact in a small area is called creep, where creep speed is defined as:

$$\text{Creep speed} = \text{traveling speed - rolling speed}$$

Fig. 1.10 Two-axle bogie model. *Source* Matsudaira, T., "Basic Theoretical Consideration on Hunting Motion of 2-axle Bogie", Paper no. 63 at the third study meeting, 1947

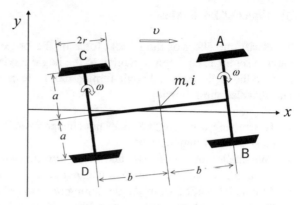

The creep force generated at the contact point due to the creep phenomenon is expressed as

$$\text{Creep force} = -f \times \text{creep speed} / \text{rolling speed},\ \text{where rolling speed}$$
$$= \text{angular velocity} \times \text{wheel radius}.$$

The proportionality constant f is called the creep coefficient.

For each wheel, Matsudaira described the speed in the wheel rotation direction, the speed in the axle direction, the wheel radius at the contact point with the rail, the rolling speed, the creep force in the wheel rotation direction, and the creep force in the axle direction. He then set up an equation of motion for the force and rotational moment acting on the bogie, solved them, and found the hunting wavelength S_2 of the two-axle bogie as follows:

$$S_2 = S_1 \times \sqrt{\left(1 + \frac{f_2 b^2}{f_1 a^2}\right)} \tag{1.2}$$

where S_1 is the wavelength of the geometric hunting, and f_1 and f_2 are the creep coefficients in the traveling and lateral directions, respectively.

Matsudaira calculated from the preceding equation that the type TR25 bogie's hunting wavelength would be 35.1 m. However, in reality it was known to be 10 to 15 m. He acknowledged this discrepancy and commented at the meeting that there would be many possible causes. Still, the most important one seemed to be the lateral elasticity and rattling between the axle and the bogie frame. That is, the lateral elasticity and play reduce the effect of the bogie's restraint on the axles, allowing the axles to move more like the hunting motion of a single wheelset. This view was confirmed in the test conducted the following year.

(3) **Effect of Bogie Mass**

To clarify the effect of the inertial force of the bogie mass on the hunting, Matsudaira set up an equation of motion of the bogie model shown in Fig. 1.10 (where m, i are the bogie mass and bogie turning radius, respectively), solved the equation, and revealed the following:

- The hunting wavelength S_2 does not change even if the inertia force of the bogie mass is taken into account.
- The smaller the bogie mass and bogie turning radius, and the larger the lateral creep coefficient and hunting wavelength S_2, the smaller the instability.
- Although the effect of the bogie turning radius is not significant, the longitudinal to lateral ratio b /a of the bogie in Fig. 1.10 has a significant impact on stability because it changes the hunting wavelength S_2.

(4) **Effect of Elastically Supported Bodies**

Matsudaira used Fig. 1.11 to examine this problem.

First, if the bogie moves very slowly, then, since the hunting motion is very slow , the bogie mass's inertia is small, so the body should move with the bogie. Therefore, this case is simply the same as an increase in the bogie mass, which results in unstable vibration.

Next, consider a case of sufficiently high speed. If hunting occurs, the vibration frequency is high; therefore, the inertia of the car body is so high that the car body should hardly move. In this case, the bogie would be stable since it is almost the same as connected to a stationary point by a spring. However, if the bogie's inertia force cannot be suppressed by the spring, the bogie becomes unstable again. The following equation can approximate the critical speed below and above the stable range.

Fig. 1.11 Model of elastically supported body. *Source* Matsudaira, T., "Basic Theoretical Consideration on Hunting Motion of 2-axle Bogie", Paper no. 63 at the third study meeting, 1947

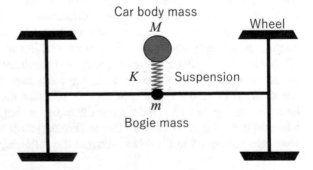

Fig. 1.12 Analytical model for the case where axles are flexibly coupled to bogie frame. *Source* Matsudaira, T., "Basic Theoretical Consideration on Hunting Motion of 2-axle Bogie", Paper no. 63 at the third study meeting, 1947

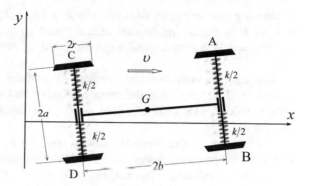

The lower critical speed:

$$S_2 \cdot n_M \tag{1.3}$$

where n_M is the resonance frequency of the body mass M and the spring constant K.
   The upper critical speed:

$$S_2 \cdot n_{Mm} \tag{1.4}$$

where n_{Mm} is the resonance frequency of the car body mass M, spring constant K, and bogie mass m. S_2 is given by Eq. 1.2.

(5) Effect of the Wheel Axle Being Flexibly Coupled to the Bogie Frame

As a significant case, Matsudaira analyzed the effect of the elasticity between the axles and the bogie frame on the hunting motion using a model shown in Fig. 1.12. Regarding the need for this analysis, Matsudaira said [4]:

> In conventional studies of hunting motion, for example by Carter,[5] the axles were fixed to the bogie frame. In an ordinary bogie, however, wheelsets can undergo considerable lateral displacement relative to bogie frames. This is caused by lateral gaps between axles and bearings, bearings and axle boxes, axle boxes and axle box guides, as well as by elastic deformation of bogie frames. So, I will investigate how the relative displacement of axles to bogie frames affects the hunting motion. For the sake of analysis, it is assumed that this relative displacement is linearly elastic. (Author's translation of the Japanese)

He developed equations of motion for each axle and the bogie, taking into account the creep force acting on each wheel, and approximately solved them, with the following results.

---

[5] F. Carter, "On the stability of running locomotives," *Proc. Roy. Soc.*, 1928.

- Hunting motion can be suppressed by k in Fig. 1.12.
- · As k increases, the hunting critical speed v_c at which hunting begins also increases. However, beyond a specific value, v_c becomes smaller.

Although the equations are complex, Matsudaira described v_c as a function consisting of a, b, k, r, and the creep coefficient and showed that for TR25-type bogies, a k of 315 kgf/mm would result in a hunting critical speed of about 200 km/h.

The fact that k was found to be very effective in suppressing hunting, albeit within the scope of the theory, seems to have convinced Matsudaira that he could control this troublesome vibration. His study was the first mathematical model of a bogie where the axles are flexibly coupled to the bogie frame, which was a crucial step in understanding the railway vehicles' running stability. However, his paper was written in Japanese and was not disseminated in the West. Meanwhile , along with theoretical research, experiments to verify his theory began in earnest.

Mamoru Hosaka,[6] who moved from the Navy to RTRI in 1945, set about to verify Matsudaira's hunting theory with handmade experimental apparatus. His apparatus used 15×20 mm aluminum angles for the rails forming a track gauge of 80 mm and flangeless cones with tread slopes of 1/10 and 1/20 for the wheels. He tested the effect of changing the aspect ratio of the bogie and confirmed that Eqs. 1.1 and 1.2 were correct.

Meanwhile, Matsudaira was working on producing a roller-type experimental device using two rotating wheels as the rails. From Hosaka's experiments, Matsudaira probably realized the limitation of the method of moving a vehicle model. He must have wondered if there could be a practical device that fixes a vehicle model and moves the rails, just like a wind-tunnel test which fixes a model and moves air.

Research on hunting took the next step with a roller rig that could take a wide range of running speeds and observe the bogie behavior in detail. The roller rig started with a simple equipment to drive one- wheelset model and soon became the one for two-axle freight car model, followed by the one for two-bogie-four-axle 1/10 model and then for a 1/5 model. In 1959, it finally developed into the world's first test stand for high-speed rolling tests of actual railcars at real speed. This test stand had been planned even before the Shinkansen project started. Thanks to the advanced efforts of Matsudaira and others, it was completed in 1959 and contributed greatly to the opening of the Shinkansen in 1964. The roller rig that Matsudaira devised played a major role in the advancement of rolling stock technology around the world.

---

[6] Mamoru Hosaka joined the Imperial Japanese Navy in 1942, moved to RTRI in 1945, studied at MIT for one year in 1952, and later developed a train seat reservation system, the first online computer system in Japan. He subsequently became a professor at the University of Tokyo and the chairman of the Information Processing Society of Japan.

Fig. 1.13 Speed versus lateral vibration frequency of MOHA31-type electric railcar. *Source* Kunieda, M., "Measurement of Train Vibration", Paper no. 101 at the fourth study meeting, 1948

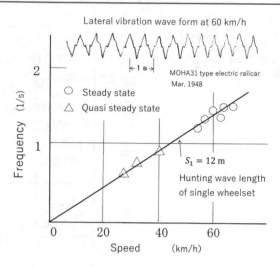

1.1.4   The Fourth Meeting

a. **Measuring the Vibration of Running Vehicle**

In February 1948, traveling vehicle's vibrations were measured for the first time after the war. A year previous, a vibration test was conducted on a MOHA63 series railcar at rest. In the tests done in 1948, vibration acceleration was measured for various types of recent railcars.

Kunieda[7] describes the test results as follows [6]:

The lateral vibration record of MOHA31 (Fig. 1.13) shows a clear wavelength of one-axle hunting, 12 m. This is probably because one-axle hunting occurs instead of two-axle hunting, as there is considerable play between the axles and the bogie frame, and as the bogie frame's rigidity is very weak. Moreover, since the wheel tread profiles would have been considerably worn to cause steep tread slopes, the flanges impacted the rails at high speed, resulting in this wavelength.

However, if bogie frames are made strong enough without giving axles any play at all, bogies will be strongly affected by rail irregularities, giving car bodies sudden changes in acceleration, which will not only make the ride uncomfortable, but will also wear down rails and wheel flanges, giving large impacts between axles and axle boxes or between axle boxes and axle box guides and resulting in failure in these parts. This has actually happened often with MOHA63 cars and others.

Based on the above considerations, it seems that a method in which axles are elastically connected to strong bogie frames without giving any play to axles is a very good solution because it prevents hunting and reduces the impact due to rail irregularities.

---

[7] Masaharu Kunieda joined the Ministry of Transportation in 1946. After serving as a senior researcher at RTRI, he later became the head of the vehicle dynamics laboratory of RTRI and the chairman of the Japan Society of Mechanical Engineers.

Fig. 1.14 Riding quality in 1948. *Source* Kunieda, M., "Measurement of Train Vibration", Paper no. 101 at the fourth study meeting, 1948

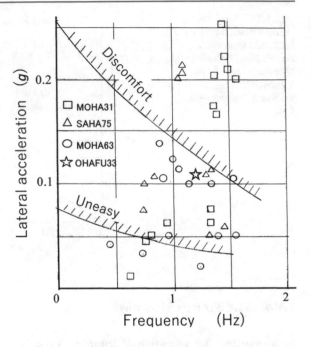

However, even with all these measures, it is impossible to escape from large rail irregularities. Since being entirely irregular, such irregularities give car bodies forced vibrations and displacement of all cycles at low speeds, and, at high speeds, shockingly act on car bodies to cause free vibrations.

Therefore, it is not enough to adjust the natural frequencies of vehicles , but it is desirable to install good dampers, so the research for that will be an important subject for the future. (Author's translation of the Japanese)

Figure 1.14 shows the data obtained from the tests, including the ride quality index. Most of the data are above the line of concern, and many of them are even above the line of discomfort, which shows how uncomfortable the ride was.

b. **Manufacture of New Bogies**

Through the aforementioned process, three new prototype bogies were built based on the discussions in the study group. First, a trial run of TR37 bogie was conducted just before the fourth study group meeting. Matsudaira, who rode the trial run, said at the meeting that side-to-side movement was reduced from one-half to one-third of the previous TR35 bogie, no hunting movement was felt, and vertical motion was also reduced, though not as much as the lateral movement. He also reported that there was quite a bit of chattering but that it was an improvement over the TR35.

Based on these results, Matsudaira stopped deepening the vibration analysis of bogies. Perhaps, he might have thought that, given the original goal of creating comfortable cars, it was more appropriate to aim for bogies without nonlinear elements than to conduct a significantly difficult analysis by incorporating rattling, friction, or spring constants with amplitude dependence that exist throughout actual bogies.

1.1.5   The Fifth Meeting

a. Test of New Bogies

The new bogies were TR37, OK1, and MD1 from Fuso Metal Industries, Kawasaki Rolling Stock, and Mitsubishi Heavy Industries, respectively. They incorporated the recommendations of the study group, such as changing the spring stiffness ratio of primary and secondary suspension, adopting highly rigid bogie frames, using axle box support systems with less rattle, using coil springs with built-in snubbers as secondary suspension, and using physically or equivalently longer swing hangar links. Tests were conducted in March 1948 on the Tokaido Line at speeds ranging from 30 to 90 km/h on straight sections and 30 to 70 km/h on curved sections.

Kunieda summarized the results as follows:

(1)   The vibrations of the new bogies were considerably reduced compared to the old ones. In particular, the decrease in lateral movement was remarkable. However, in terms of chattering, the new bogies were hardly improved compared to the old.

(2)   To put it in more detail,

- The frequencies of vertical vibrations were 1.5 to 2.5 Hz.
- The wavelengths of lateral vibrations due to the hunting motion ranged from 10 to 20 m.
- The vibration frequencies ranged from 0.6 to 1.0 Hz at low speeds below 50 km/h, where the hunting caused the car body to roll.
- At high speeds above 50 km/h, body yawing of 1.2 to 2.0 Hz appeared.

Thus, the new bogies born from the study group cleared up the major problems that had made the ride extremely uncomfortable, but many issues that were previously invisible or low on the list emerged. The tests took the vehicle's vibration prevention to the next step.

1.1.6   The Sixth Meeting

The sixth meeting was held in April 1949. The moderator made the following opening address:

It will be a challenging year due to the budgetary constraints on research and prototype production because of the need to rebuild the Japanese economy according to the Nine Principles.[8] The Ministry of Transport has conducted various tests of prototype bogies, but it is feared that few major tests will be possible this year. (Author's translation of the Japanese)

In the sixth meeting, Matsudaira Group presented two interesting reports. The first one was titled "Experimental Study on Hunting Motion of a Scale Model Bogie," in which the hunting theory presented at the third meeting was verified by using a roller rig, and the second one was titled "Car Body Vibration Characteristics when Viscous Damping Is Added to Secondary Suspension."

Matsudaira's first analysis of vehicle's natural frequencies was based on the assumption that there was no friction in suspension. Kunieda then discussed the case where the suspension had a constant frictional force, followed by the case where it had Coulomb frictional force proportional to loads. This time, Matsudaira analyzed the case where resistance is proportional to velocity. In this paper, Matsudaira said that chattering vibrations cannot be removed by frictional devices; they must be removed by oil dampers. That is, the only way to prevent transmission of chattering vibrations from bogie frames to car bodies is to have frictionless coil springs embrace oil dampers whose resistance is proportional to speeds (= vibration amplitude × vibration frequency), thus blocking high frequencies.

At the end of the meeting, many research topics that should be studied in the future were proposed, including the following: deepening the hunting study by using model cars and roller rigs, developing support methods for main motors, designing bogies using oil dampers, using air springs, using anti-vibration rubber, and evaluating the necessity/unnecessity of swing hanger systems.

The next meeting was to be held in September of that year, but this sixth meeting was the last. In March, a month before the sixth meeting, GHQ presented the Japanese government with a budget plan for fiscal 1949 that included implementing the nine principles. Based on the budget plan, a cabinet decision was made on May 4 to reorganize the administrative organizations, the Act on the Capacity of Administrative Organizations was enacted on the May 30, and the railway division was separated from the Ministry of Transport on June 1, thereby establishing the Japanese National Railways (JNR). Subsequently, on June 24, JNR labor union decided to use its power to oppose administrative reorganization; on July 6, JNR President Shimoyama was run over and killed by a train; on July 12, JNR notified the labor union of the reduction of 63,000 employees. Then, a crewless train runaway incident occurred on July 15, and a passenger train derailment and an overturning incident occurred on August 17. In short, a series of mysterious and disturbing incidents occurred. One year later, on June 25, 1950, the Korean War broke out.

---

[8] The nine principles were the economic policies aimed at stabilizing the postwar Japanese economy; they were ordered by the General Headquarters (GHQ) on December 19, 1948.

When the sixth meeting was held on April 20–21, 1949, it was probably still calm before a storm. However, as days went by, conditions were no longer conducive to continuing the study group's activity.

In analyzing the study group's progress, it became clear that the fact that study group started in 1946 had a close relationship with the commencement of the Shinkansen's operation in 1964. The direction of the fundamental issues of vehicle vibrations and hunting motion had become clear by the sixth meeting.

If the start of the study group had been delayed for a year, the study group members would have been forced to stop activities without producing any results due to the nine principles. That is, they would not have been able to make new prototypes, nor would they have been able to test them on the track. In that case, the RTRI's commemorative lecture held in May 1957, entitled "The Possibility of a Three-Hour Tokyo–Osaka Service" (see Chap. 2), would not have been realized, and the Shinkansen project would not have taken shape.

1.1.7   From 1949 to 1957

a. **Birth of Series 80 Train**

The study meeting ended in April 1949, but soon a new railcar was born based on its results. In March 1950, the Series 80 railcar with DT16-type bogie appeared. DT16 bogie used a cast steel frame to increase its rigidity, soft primary suspension to increase the amount of extension and contraction, and lengthened swing hangar links to improve riding comfort. However, the study group's results had not been fully incorporated because layered leaf springs were still used for the secondary suspension (Photo 1.4).

In 1952, DT16 evolved into DT17, which adopted oil dampers and coil springs instead of layered leaf springs for the secondary suspension. Since the leaf springs were gone, the vibration system's characteristics were recalculated , and the riding comfort of the Series 80 trains with the primary and secondary suspension's stiffness and oil dampers set to the optimum values was greatly improved. The Series 80 trains operated 126 km from Tokyo to Numazu on the Tokaido Line in two and

Photo 1.4 Series 80 electric train. (provided by Tatsuya Miyasaka)

a half hours, demonstrating that long-distance electric railcar trains were possible. The study group met Shima's expectation.

b. **Verification of Hunting Theory of a Single Wheelset by Scale Model Experiments**

Verification of hunting theory began in earnest by using a scale model and a roller rig. Matsudaira generalized the analysis of a single-wheelset hunting motion by using an analysis model in which the axle is elastically coupled to the bogie frame, as shown in Fig. 1.16. Figure 1.15 shows the roller rig used. The apparatus had a roller diameter of 500 mm, a gauge of 80 mm, wheel diameters of 40, 60, and 80 mm, and tread slopes of 1/20 and 1/10.

Although there are no records of the process of building the experimental system, Matsudaira group likely repeated trial and error many times to obtain reliable data on a 1/10 scale model, including variable speed transmission, wheelset retention, and hunting motion recording mechanism. Matsudaira described the purpose of the experiment as follows [7]:

> The experiment has two purposes. The first experiment is to clarify that the hunting motion of a wheelset is mechanically unstable motion, and to confirm the theoretical formula of the hunting wavelength. The second is to investigate the effect of elastic restoring force of the axle on the hunting motion. (Author's translation of the Japanese)

Fig. 1.15 Single-wheelset roller rig with 1/10 scale wheelset model. *Source* Matsudaira, T., "Shimmy of Axle of a Pair of Wheels", RTRI Research Materials, no.19, RTRI, 1952, p. 24

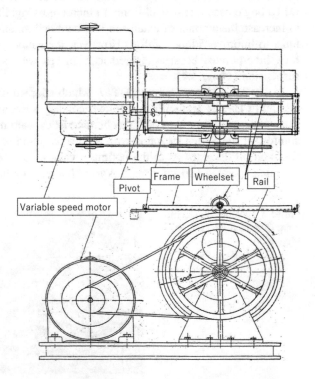

Fig. 1.16 Analysis model for a wheelset elastically coupled to bogie frame. *Source* Matsudaira, T., "Shimmy of Axle of a Pair of Wheels", RTRI Research Materials, no.19, RTRI, 1952, p. 21

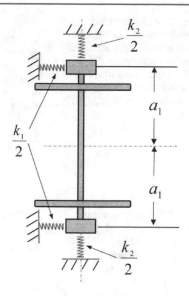

Figure 1.17 shows a comparison of the theoretical and experimental values of the hunting wavelength. The fact that the hunting occurs at all speeds shows that hunting is essentially an unstable oscillation and the fact that the hunting frequency coincides with the value calculated by Eq. 1.1 indicates that Eq. 1.1 is correct. The figure shows the results for a wheel diameter of 60 mm, but similar results were obtained for 40 mm and 80 mm wheels.

Matsudaira's analysis showed that the hunting frequency of the model in Fig. 1.16 is a linear combination of the lateral natural frequency and yawing natural frequency of the wheelset and showed that the hunting frequency is proportional to the yawing frequency when $k_2=0$. Fig. 1.18 shows a comparison of the calculated and experimental values for the hunting frequency and yawing natural frequency when $k_2=0$, indicating that his analysis was correct.

Matsudaira conducted more than 150 experiments to verify the geometric hunting wavelengths shown in Fig. 1.17. Of course, such high-density demonstration tests are impossible with an actual vehicle and even with running-type model apparatus. His experiments show the power of the rolling rig method.

c. **Measures against Hunting Motion of Freight Wagons**

As described previously, the hunting motion had been studied as a factor that worsened train ride quality. Nonetheless, the study results were instrumental in increasing the speed of freight trains.

Matsudaira described the hunting motion of freight wagons as follows [8]:

Fig. 1.17 Relationship
between speed and hunting
motion of a single wheelset.
Source Matsudaira, T.,
"Shimmy of Axle of a Pair of
Wheels", RTRI Research
Materials, no.19, RTRI, 1952,
p. 25

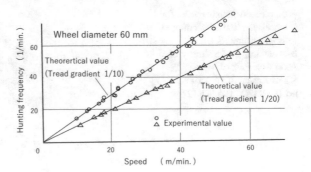

Fig. 1.18 Yawing natural
frequency and hunting critical
frequency when $k_2 = 0$.
Source Matsudaira, T.,
"Shimmy of Axle of a Pair of
Wheels", RTRI Research
Materials, no.19, RTRI, 1952,
p. 26

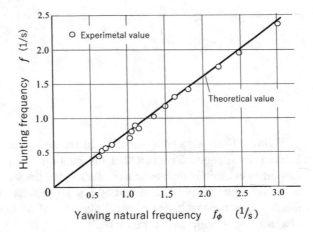

Currently, the maximum permissible speed of two-axle wagons in Japan is set at 65 km/h. The basis for this figure's determination is not clear but is probably mainly empirical and related to their derailment accidents. As has been shown by numerous experiments, current two-axle wagons generally tend to experience a sudden increase in lateral motion when speeds exceed 50–60 km/h, especially when wheel treads are severely worn.

This lateral vibration is a kind of self-excited oscillation called "hunting motion" and is not primarily related to track's quality.

Almost all derailments of two-axle wagons during high-speed operations are thought to be caused primarily by severe lateral movement of the vehicle due to this meandering, which is triggered by localized defects in tracks. (Author's translation of the Japanese)

Matsudaira modeled two-axle wagons, as shown in Fig. 1.19, and performed hunting analysis. However, because the order of the equations was too high to solve analytically, he had to spend a lot of time and effort obtaining the solution by

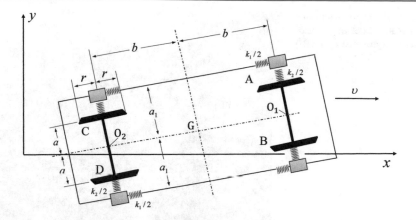

Fig. 1.19 Analysis model for two-axle freight wagons. *Source* Matsudaira, T. et al., "Train Speed-up through the Improvement of Spring Suspension System of 2-Axle Freight Cars", RTRI Research Materials, no.18, RTRI, 1953, p. 5

graphical methods. (In those days, there were no computers.) He thought it was important that the axles were tightly coupled to the bogie frame in the traveling direction and elastically coupled in the lateral direction, so he set k_1 in Fig. 1.19 to infinity and found the relationship between k_2 and hunting critical speed. Figure 1.20 shows the results.

The figure shows two unstable regions where hunting occurs and two stable regions where hunting converges. For example, in a wagon with lateral support stiffness of k_a, hunting occurs when the speed reaches v_a and does not settle. On the other hand, in a wagon with lateral support rigidity of k_b, hunting occurs at a speed

Fig. 1.20 Lateral elasticity of axle coupling and hunting critical speed. *Source* Matsudaira, T. et al., "Train Speed-up through the Improvement of Spring Suspension System of 2-Axle Freight Cars", RTRI Research Materials, no.18, RTRI, 1953, p. 5

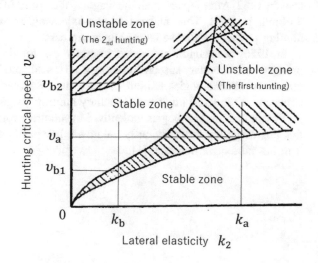

Photo 1.5 Experiment using 1/10 scale model wagon and roller rig (provided by RTRI)

of near v_{b1}, but when the speed increases a little, the vibration enters a stable region and the hunting stops. After that, no hunting will occur until the speed reaches v_{b2}. The effect of weakening the lateral support rigidity of the axles of the freight wagons at the time, which had the rigidity of k_a, to k_b was confirmed in running tests in 1952. After repairing all the wagons, the speed of freight trains was raised to 75 km/h in 1968. This analysis of two-axle wagons was a prelude to studying hunting prevention for the later Shinkansen cars.

In 1956, Matsudaira proceeded to study the hunting motion of railcars with primary and secondary suspension using a 1/10 scale model and roller rig and found that conventional bogies exhibit primary hunting (car body hunting) at speeds between 100 and 150 km/h and secondary hunting (bogie hunting) at speeds above 200 km/h that vibrates bogies violently. Matsudaira wrote in his memoirs that since this bogie hunting was previously unknown, he specifically focused on controlling it in his subsequent research [1].

1.2 Strength, Weight Reduction, and Aerodynamic Drag of Car Bodies

After graduating from university in 1933, Lieutenant Colonel Tadanao Miki,[9] who had designed aircraft at the Naval Aviation Technology Center, started working at the Railway Technical Research Institute in December 1945. There, he utilized the technology he had developed for the bomber "Ginga" (which means the galaxy) and other aircraft to realize high-speed railcars.

In his memoirs, Miki wrote [9]:

> As the war situation became unfavorable, aircraft's air superiority diminished, and pilots' skills declined. A so-called suicide mission was devised by front lines and submitted to aeronautical engineers. It was a plan for human bombs. Despite our engineers' opposition, the plan was formally taken up, and a prototype order was issued. The test flight met all the requirements, and it was named the "Ouka" (which means cherry blossom). War is always a close call with death, and it was heartbreaking for me to think of the many young men who died in the "Ouka."

> The horrors of war never left my mind, and after the war, I decided not to work in jobs related to the war. So, I joined the Railway Technology Research Institute and began researching light-weight high-speed railcars using the technology I had developed in the airplane field. (Author's translation of the Japanese) (Photo 1.6)

At the time, those who moved from the military and other organizations to the RTRI were assigned to the Kunitachi branch office on Tokyo's outskirts, where the current RTRI main office is located (Photo 1.7).

Miki must have been surprised at the branch office's poor research environment. He did not write about it, but Matsudaira, who also moved from the Naval Aviation Technology Center, wrote in his memoirs about the branch office's conditions as follows [10]: (Photo 1.7)

> As a long-time employee of the Naval Air Technology Center, probably one of the world's leading research facilities at the time, I expected the Kunitachi branch would have adequate facilities. Still, all that stood before me were a couple of shabby wooden barracks, which did not look like laboratories by all appearances. However, as it read "Railway Technical Research Institute' on a small gatepost that stood near the entrance, there was no mistake. (Author's translation of the Japanese)

The fact that they began their railway research in such a poor environment and achieved the world's first high-speed railway in less than 20 years is a testament to their abilities.

1.2.1 Realization of Stress Measurement by Strain Gauges

Wire strain gauges had been used in the USA since the end of World War II, but only a little information was available in Japan.

---

[9] Tadanao Miki joined the Imperial Japanese Navy in 1933, moved to RTRI in 1945 as a senior researcher, and later became the head of the vehicle structure laboratory of RTRI.

Photo 1.6 Tadanao Miki (provided by RTRI)

Photo 1.7 The then Kunitachi branch (provided by RTRI)

Figure 1.21 shows an example of strain gauges currently in use. It is made very thin so as to be attached to base material. As the base material extends, the wire of the strain gauge extends and becomes thinner, and the electrical resistance increases, so by measuring the electrical resistance of the gauge, the force applied to the base material can be determined.

Photo 1.8 The then RTRI main office (provided by RTRI)

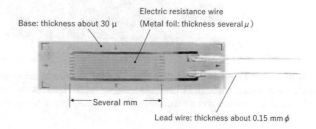

Fig. 1.21 Current strain gauge

Japanese strain gauges were developed by Nakamura[10] and others. Nakamura handmade them using a 0.02 -mm-diameter resistive wire, as shown in Fig. 1.22. Nakae, who was Nakamura's assistant, said [11]:

---

[10] Kazuo Nakamura joined the Imperial Japanese Army in 1942 and moved to RTRI from Nakajima Aircraft Manufacturing Co. Ltd. in 1946 as a senior researcher.

Fig. 1.22 Handmade strain gauge. *Source* Nakae, S., "Exploring the Roots of Strain Gauge", Railway Research Review, no.1, RTRI, 2009, p. 1

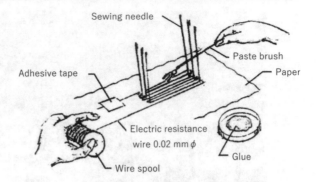

Photo 1.9 Stress measurement of a car body using strain gauges (provided by RTRI)

At that time, it was tough to gather information and material because of the confusion after the war. Paper gauges were completed in 1949. At first, we made them for our own use, but some people in other laboratories and even outside the laboratory began to ask to use them . (Author's translation of the Japanese)

In 1950, Nakamura succeeded in simultaneously measuring strains at multiple locations in vehicle load tests, thereby paving the way for vehicle weight reduction (Photo 1.9).

Strain gauges made it possible to measure dynamic wheel loads and lateral force upon rails, stresses of wheel axles, impact force acting on automatic couplers, stresses on each part of car bodies, vibration acceleration at each part of tracks, and so on, helping many technologies open new phases of research based on supporting data. It can be said that Nakamura's strain gauge played a major role in the development of railway technology. Strain gauges became commercially available in 1952.

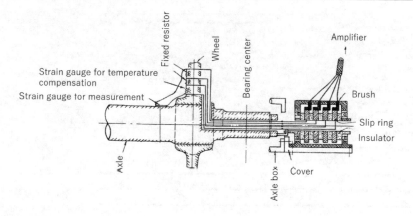

Fig. 1.23 Wheel shaft strain measuring device. *Source* Research on High-Speed Railways, Kenyusha Foundation, 1967, p. 210

1.2.2 Measurement of Dynamic Wheel Load and Lateral Force upon Rails

The development of the strain gauge was initially started to study the dynamic problems of wheelsets. However, its application on wheelsets was interrupted because priorities were placed on measuring car body stress, so it was not until December 1954 that a device to measure rotating wheelsets' strain was developed.

A set of slip rings shown in Fig. 1.23, which transmits signals from the rotating parts to the stationary portions, was the device's technological key.

The output of the strain gauge is extremely weak, so the gauge cannot be used if noise is generated during signal transmitting. As a result of the know-how obtained in the development process, such as the best ring and brushes' material and the way of holding brushes in place, the slip ring's contact noise became sufficiently low compared to the signal level and allowed accurate measurement of wheel and axle strain.

With this apparatus, the first measurement of the dynamic stresses on a wheel and axle while running was made in July 1954 (Fig. 1.24).

Nakamura wrote in his paper [12]:

The results were extremely good, with wheel loads due to running and impacting stresses at turnout points well recorded, indicating that the slip ring had adequate performance. The measurements showed that the maximum lateral force on the wheel was 2,300 kgf (0.4 times the static wheel load) and the maximum wheel load was 1.4 times the static load. (Author's translation of the Japanese)

Thereafter, this device provided data that would ensure the safety of the ride.

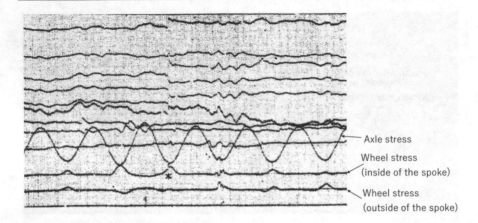

Axle stress

Wheel stress
(inside of the spoke)

Wheel stress
(outside of the spoke)

Fig. 1.24 First axle and wheel stress measurements (1954). *Source* Nakamura, K. et al., "Dynamic Problems in Railway Wheelsets", RTRI Research Materials, no. 11, RTRI, 1956, p. 8

1.2.3   Weight Reduction of Railway Vehicles

Figure 1.25 shows the weight breakdown of a MOHA80-type railcar before weight reduction (Photo 1.4). As shown in the figure, it is clear that the overall weight reduction requires that not only the car body but also many other elements are reduced in weight.

a. Car Body Frames

The body has six frameworks, an underframe, side frames, front and rear frames, and a roof framework. The floor, side panels, and roof are attached to them. In the days when a wooden body was mounted on a steel underframe, the underframe was made to provide all the strength, but when the wooden parts were replaced by steel, it became possible to provide the load to these as well.

Since the vehicle body receives force such as its weight, load, twist, and longitudinal compression/tension, it must be strong enough not to be damaged by these. Even if it is not damaged, if deformation is too large, it is unacceptable from the viewpoint of riding comfort. Thus, the test method to confirm both strength and stiffness was completed in 1953 by the practical application of strain gauges.

For example, for the vertical bending strength, weights were placed on the floor in a range of 0 to 20 tons, and for torsional strength, one end was twisted with force ranging from 0 to 4.8 tf with another end fixed. It was also determined that stresses should be measured at 200 locations using strain gauges (Fig. 1.26).

Figure 1.26 shows the bending of a car body in a load test. With this body, a distributed weight of 20 tons on the floor lowers the center of the floor by 4 mm, but when the body is compressed longitudinally by 100 tf in this condition, the floor center rises by 4 mm.

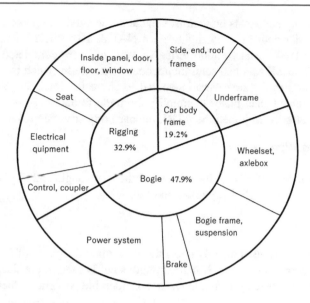

Fig. 1.25 Weight breakdown of a MOHA80-type railcar (gross 47.5 tons). *Source* Miki, T., "Trends in Weight Reduction of Railway Vehicles", The Engineering Journal for Transportation, no. 9, Kotsu Kyouryokukai, 1953, p. 26

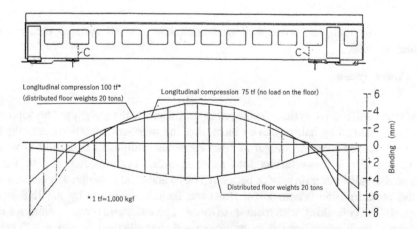

Fig. 1.26 Vertical bending strength and longitudinal compressive strength test. *Source* Miki, T., "Load test of NAHA10 Series Lightweight Passenger Car", The Engineering Journal for Transportation, no. 1, Kotsu Kyouryokukai, 1956, p. 39

By conducting such tests on various vehicle bodies, the margin of each member strength was clarified, in 1955, lightweight NAHA 10 vehicle bodies were produced, and in 1957, Odakyu 3000 series SE vehicles were completed.

NAHA10 and SE cars had semi-monocoque structure in which the underframe, side frames, and roof frame were integrated to share the load, and the floor was made of corrugated steel plates to simplify the underframe.

Ten years after Miki joined the RTRI, the car body structure entered a new era.

b. **Bogie and Rigging**

Figure 1.25 shows that 80% of the railcar weight is the bogie and rigging. As a way to reduce this weight, Miki suggested the following:

(1) *Bogie*

Bogie frame: Wheelbases directly affect the strength and rigidity of the frames and should be as short as possible. If high-strength steel is used in consideration of the development of welding technology, a considerable weight reduction can be expected.

Axle: If wheel mating parts are induction hardened and axles is made hollow, weight can be reduced without reducing strength and rigidity. In addition, the weight of wheels can be reduced by integrating the rim and tire. A prototype hollow axle with integrated wheel is 32% lighter than the conventional wheelset.

(2) *Rigging*

If light alloys and plastics are used, seats, upholstery, water systems, etc., can be lighter by 30%.

(3) *Power System*

With the cardan drive system developed in recent years, the weight per horsepower has been reduced by half or less by increasing the number of rotations per minute. Research on reducing the weight of other electrical equipment is desired as well.

Taking all of these factors into account, Miki put together a plan for a four-and-a-half-hour train service between Tokyo and Osaka, which was announced to the press in 1953 (Photo 1.10). However, he was criticized by the JNR head office for not consulting with them in advance. The head office was waiting for the Tokaido Line's electrification to be completed and planned to run a 120-km/h express train.

Photo 1.10 Newspaper
article detailing Miki's plan
(*The Asahi Shimbun*, Oct.17,
1953)

1.2.4   Tokyo–Osaka 4.5-h Plan

In 1953, it took eight hours to get from Tokyo to Osaka with the limited express
train Tsubame (which means a swallow) pulled by a steam locomotive running at a
maximum speed of 95 km/h. So, Miki's idea was a remarkable one (Fig. 1.27).

Miki contributed an article to a railway magazine entitled "A Plan for Super
Express Train (Tokyo–Osaka 4 and a Half Hours)." In the preface of the article, he
stated [13]:

Fig. 1.27   Tokaido Line (Tokyo–Osaka)

When the entire Tokaido Line is electrified, there is talk of running a six-and-a-half-hour express train between Tokyo and Osaka, but I wonder if this is the limit for narrow gauge trains. So, I would like to introduce an idea of a revolutionary speed increase that I have been working on for some time now, which I believe has great potential for realization. (Author's translation of the Japanese)

Miki introduced the technical details of the plan as follows:

(a) *Streamlining*: Since the power required increases rapidly as the speed increases, the head of the lead car shall be almost completely streamlined, and the tail car shall also be streamlined to minimize eddy current drag. Wind-tunnel tests have shown that the drag coefficient can be reduced by 50% compared to the conventional type by streamlining. That is, by smoothing the roofs, carriage windows, and couplers, covering the lower part of the vehicles, and reducing the cross-sectional area as shown in Fig. 1.28, the aerodynamic drag can be reduced to less than half of the current value.

(b) *Weight reduction*: The body shall be made of corrosion-resistant light alloy, and the structure shall have a stressed skin structure with high bending and torsional rigidity. To reduce weight and noise, the interior shall be upholstered with non-flammable fabric similar to that used in airplanes and automobiles. Employing light alloy for the chairs, as in airplanes, will reduce the current weight of 118 kg to about 25 kg. The water system shall be made of glass fiber resin and light alloy, and the bogies will be articulated to reduce weight. The axles shall be hollow shafts, the wheels be pressed plate wheels, and the bogie frames be made of high-tensile-strength steel sheet welded together or light alloy. In this way, the body

Fig. 1.28 Downsizing the car body and lowering the center of gravity. *Source* Miki, T., "A Concept for the Super Express (4.5 hours between Tokyo and Osaka)", The Engineering Journal for Transportation, no. 1, Kotsu Kyourokukai, 1954, p. 3

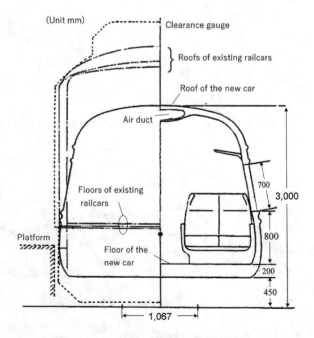

is estimated at 3 tons, the bogie at 3.7 tons, and the outfitting (including the water system and air conditioning) at 4.75 tons, resulting in 11.45 tons per car for a 32-passenger car.

(c) *Lowering the center of gravity*: The height of the center of gravity of 1,450 mm in the current passenger coaches will be reduced to 1,080 mm.

(d) *Reinforced brakes:* Disk brakes will be adopted. At the time of emergency braking from the highest speed, it would be advantageous to reduce the braking distance if air resistance plates are installed at the locomotive head, at the end sides of the rear-most car, and on the roof to the extent that the vehicle's limits allow.

(e) *Locomotive*: Electric railcars have to mount electric equipment under the floor, so they cannot be a low floor type. The locomotive system is advantageous from the viewpoint of air resistance and safety against crosswind by lowering the center of gravity. The motive power system is electric or diesel, but electricity will be good considering the Tokaido Line's electrification. The required power will be 1,000 kW (1,350 horsepower). The front of the locomotive shall be entirely streamlined, the equipment shall be arranged so that the cross section becomes as small as possible, the roof shall be round, and the center of gravity shall be low, similar to passenger cars.

Miki, an aircraft designer, designed a modern seven-car train pulled by a locomotive, with a maximum of 200 passengers and a maximum speed of 160 km/h, that would connect Tokyo and Osaka in 4 h and 45 min, including 12 min of stopping time at intermediate stations.

This concept, however, was never realized by the Japanese National Railways.

1.2.5 Odakyu's Type 3000 SE Train

Although there was considerable criticism from JNR head office, the situation took an unexpected turn. The Ministry of Transportation took notice of Miki's plan and notified him of a research grant. Accepting the grant, Miki began to research to bring his concept to fruition. He wrote in his memoirs [9]:

> I initially considered an electric locomotive train, but I decided to use a power decentralized train system in anticipation of future improvements in electric equipment and drive systems.

> … Air resistance is proportional to the square of the speed, so the ratio of air resistance to running resistance becomes very large at high speed. However, there were few examples of measuring train's air resistance in Japan, so I decided to use the research grant to conduct wind tunnel tests of scale models. (Author's translation of the Japanese)

Thus, Miki's full-scale wind-tunnel testing began (see Sect. 4.7).

In October 1954, Miki received an unexpected request from Odakyu Electric Railway Company, which had been running a 100-min service between Shinjuku and Odawara (Fig. 1.29). They were researching ways to reduce the time to 60 min

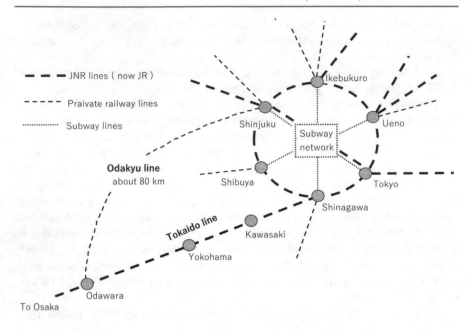

Fig. 1.29 Odakyu line (Shinjuku–Odawara)

to compete with the JNR's Tokaido Line, so the newspaper article about Miki's rapid train had caught their attention. They sought guidance and assistance from Miki in all aspects of planning and design to realize a train that would surpass the world's standards in shape and performance.

In May 1956, the specifications for the eight-car SE train (Super Express) were decided almost exactly as Miki had envisioned, and production began. The limited express train by SE cars started commercial operation in July 1957, received high praise from passengers, and contributed to the company's management.

Meanwhile, in November 1956, after completing electrification work on the Tokaido Line, JNR began operating limited express trains with electric locomotives at 95 km/h. However, since they planned to run a new train with electric railcars at 120 km/h, they needed technical data at 120 km/h or higher. Thus, high-speed testing of SE cars was conducted on the JNR's Tokaido Line in September 1957. In this test, the SE train recorded a maximum speed of 143 km/h on the narrow gauge track.

The technologies used in the SE train, such as streamlined lead cars, semi-monocoque bodies, disk brakes, lightweight seats, and aerodynamic skirts on the lead car, were carried over to the Shinkansen vehicle. Having received a research grant to develop the SE train, Miki had accumulated much knowledge relating to the Shinkansen train's design (Photo 1.11).

Photo 1.11   Odakyu 3000 SE train (provided by RTRI)

1.3   Continuous Welded Rail and Track Dynamics

1.3.1   Continuous Welded Rail (CWR)

Rail length is determined by manufacturing processes, transportation restrictions, and maintenance workability. In Japan, in 1933, the 50kg rail and 37kg rail were decided to 25 m, and the 30kg rail was decided to 20 m. Seams that connect rails are the weakest point in tracks, and it is said that the maintenance work related to seams amounts to 20 to 40% of the total volume of maintenance. In addition, seams impair riding comfort and vehicles, so seamless long rails have long been a concern for railroads, and much effort had been expended to make them happen.

CWR tracks must meet the following conditions:

(i)   Expansion or contraction of CWR that is about ±40 mm can be absorbed by expansion joints or buffer rails.

(ii)   Sufficient rail fastening force and ballast resistance must be provided so that rails do not slide or tracks do not move forward due to the axial pressure and tensile force generated on rails.

Fig. 1.30 Length of end part that can expand/contract with temperature change. *Source* Sato, Y., Track Dynamics, Tetsudo Gengyousha, 1964, p. 90

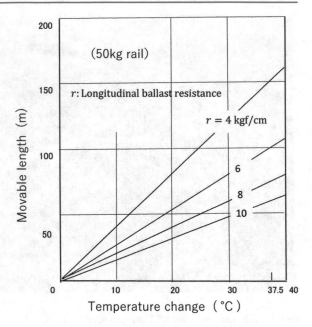

(iii)   Welded joints shall not break under the large tensile force generated on rails in severe winter weather.

(iv)   There must be sufficient lateral resistance and rigidity to prevent rails from buckling due to the large axial force generated in the rails in high temperature of summer.

a. **Expansion and Contraction of CWR Tracks**

Rails expand or contract according to changes in temperature, but when they are attached to sleepers, they cannot expand or contract freely due to the sleeper's resistance. Suppose the sleeper's resistance is sufficiently high. In that case, the expansion or contraction of the central section of the rails will be suppressed by the sleeper's resistance so that only the end part of the rails will expand or contract no matter how long the rail is.

According to Hoshino's[11] analysis, the length of the end part that can expand or contract due to temperature change is as shown in Fig. 1.30.

Figure 1.30 shows that the length of the end part that can stretch or retract is about 75 m, even with a temperature change of 37 °C when the ballast resistance in the rail direction is 8 kgf/cm. In this case, what is the amount of expansion or

---

[11] Yoichi Hoshino joined the Ministry of Railways in 1931. He later became the head of the track laboratory of RTRI.

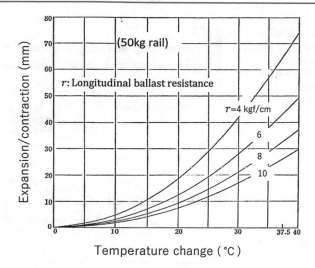

Fig. 1.31 Expansion/contraction of the rail. *Source* Sato, Y., Track Dynamics, Tetsudo Gengyousha, 1964, p. 90

contraction in that part? Figure 1.31 shows the relationship among ballast resistance, temperature change, and the amount of expansion or contraction at the rail end.

From the figure, it can be seen that in the case of a 50kg rail, the amount of expansion or contraction is only about 30 mm (15 mm per side) for a temperature change of 37 °C.

To test Hoshin's theory, a 210-m CWR test track was laid in a yard in November 1939, and the movement of the rail was measured for two years when locomotives were running on the track. In theory, the track was supposed to be safe. However, as it was the first long rail in a business line, Hoshino took anti-buckling measures to the track and placed joint plates at the welding points to ensure safety. The rail movement and axial force were measured twice a day for two years, and the analysis proved to be correct. Commenting on the results, Hoshino said [14]:

> If tracks are as well-maintained and straight as this test track, it will be possible to adopt infinite length rails. However, the temperature for laying rails should be limited to 25 to 30 degrees Celsius, which is a very restrictive temperature. In addition, we must be prepared to face considerable ballast restrictions on raking, rail cutting, and joint opening. The buckling strength of Japanese tracks may be much lower than that of the U.S., so it is problematic to refer to foreign examples directly. Meanwhile, we have only one buckling test in Japan at Omiya in 1932 [described in the next section], which is insufficient for making a proper judgment, so we need more extensive experimental data. (Author's translation of the Japanese)

Fig. 1.32 Test track by
Horikoshi. *Source* Horikoshi,
I., "On the Buckling of Rails",
Research Materials, no. 18,
Railway Minister's Office
Research Institute, 1934, p. 6

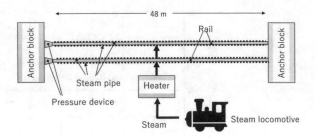

Fig. 1.33 Buckling shape.
Source Horikoshi, I., "On the
Buckling of Rails", Research
Materials, no. 18, Railway
Minister's Office Research
Institute, 1934, p. 10

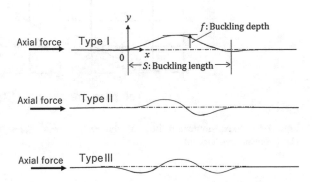

b. **Study of Track Buckling**

(1) **Step 1**

Amman and Gruenewald[12] of Germany are considered to be the pioneers of track buckling research. The primary role of sleepers and track beds has been to support train loads, but for CWR tracks, they have a new role in buckling restraint. However, since the track strength from that point of view was unknown, Horikoshi[13] temporarily installed a test facility, as shown in Fig. 1.32, in 1932 at Omiya station to investigate the track's buckling strength.

Increasing rail axial force to cause buckling was not an easy task. Horikoshi used steam from a steam locomotive and pressurizing devices, as shown in Fig. 1.32. A series of tests showed that the buckling shape could not be arbitrary but took on three different forms, as shown in Fig. 1.33 (Photo 1.12).

When the energy stored in the rails due to the increase in axial pressure reaches a specific value, buckling occurs, releasing the energy stored in the rails partially, and the track deforms and stabilizes in a particular shape, as shown in Fig. 1.33. Based on the fact that the buckling takes waveform shown in Fig. 1.33, Horikoshi derived the relationship among the axial force just before buckling, buckling length, and

---

[12] Ammann, Guenewald; Versuche uber die Wirkung von Langskraften im Gleis, *Organ,* No. 29, 1929.

[13] Ichizo Horikoshi was a then track engineer at the Railway Minister's Office Research Institute (now RTRI).

Photo 1.12 Buckling of the test track.[14]

buckling depth using the relationship among the amount of work done by the axial force during the buckling process, energy absorbed by the ballast, and strain energy stored in the bent rails. He spent a lot of labor and time to create nomograms, as shown in Fig. 1.34, by manual calculation.

Figure 1.34 shows a nomogram that reads the axial pressure, buckling length, and buckling depth when a track with 50kg rails, lateral ballast resistance of 2.04 kgf/cm, longitudinal ballast resistance of 3 kgf/cm, and radius of 500-m buckles to the type I shape shown in Fig. 1.33. It shows that the minimum axial pressure at which buckling occurs is 50.6 tf, the buckling length at that time is 12.55 m, the buckling depth is 15 cm, and if buckling does not occur at 50.6 tf but does occur at 60 tf, the buckling length is 14.75 m and the depth is 30 cm.

Horikoshi's research did not lead to long rails' practical application. The reason for this is not clear, but there was likely a lack of accurate experimental data to drive a large-scale demonstration experiment. In most of Horikoshi's experiments, the rail axial force was increased by the pressure device. This method might not correctly simulate the increase in rail axial force due to temperature rise.

[14] Horikoshi, I., "On the Buckling of Rails", *Research Materials*, no. 18, Railway Minister's Office Research Institute, 1934, p. 14.

Fig. 1.34 An example of
nomogram by Horikoshi.
Source Horikoshi, I., "On the
Buckling of Rails", Research
Materials, no. 18, Railway
Minister's Office Research
Institute, 1934, p. 38

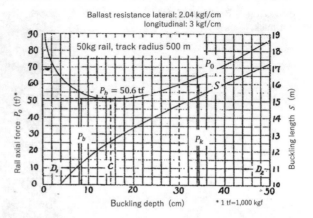

(2) Step 2

The CWR research was resumed by Minoru Numata[15] in 1953 after a long interruption of Horikoshi's work. Numata extended Horikoshi's analysis and derived
general equations that calculate the buckling length and depth for various track
conditions. He also spent a lot of time and effort to produce many nomographs like
that shown in Fig. 1.35 (remember, this was before the advent of computers).

Figure 1.35 shows the relationship among axial force, buckling depth, and
buckling length in relation to lateral ballast resistance when a track with 50kg rails
buckles to the type I shape shown in Fig. 1.33.

The challenge was in verifying these calculations. The difficulty of the experiment in causing buckling by heating real rails was apparent from Horikoshi's work
at Omiya.

Numata solved this problem by experiments using a scale model of about 1/10
and paved the way for the practical application of CWR.

Since a vehicle is a combination of metal parts, its behavior is highly repro-
ducible even in model experiments. Track's components are rails, sleepers, and
ballast, so it is expected that the accuracy of model experiments will decline.
Numata solved this problem by statistically analyzing the results of many experiments (track's buckling is a statistical phenomenon that depends on the track
deformation before buckling and the initial strain inherent in the rails).

When considering model experiments, it would be most efficient if the experimental results could be converted into real size by making the similarity rates of the
factors that control buckling the same. However, it is almost impossible to set
precisely the same similarity rates for rail stiffness and ballast resistance . Therefore,
based on the fact that theoretical formulas were already obtained, Numata adopted a

---

[15] Minoru Numata joined the Ministry of Railway in 1948. After serving as a senior researcher at
RTRI, he became a professor at Kyushu University.

Fig. 1.35 A nomograph for a straight track with 50kg rails. *Source* Numata, M., "Buckling Strength of Long Welded Rail", Railway Technical Research Report, no. 721, RTRI, 1970, p. 23

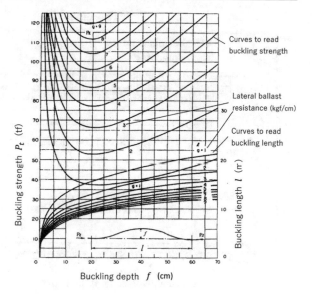

Fig. 1.36 Rail models. *Source* Numata, M., "Buckling Strength of Long Welded Rail", Railway Technical Research Report, no. 721, RTRI, 1970, p. 61

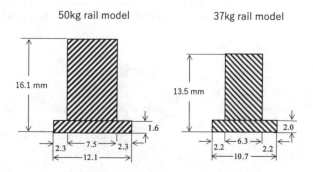

method of confirming the theoretical formulas' suitability by changing these similarity rates within the range where the reliability of the experiment was maintained. Figure 1.36 shows the rail models, and Fig. 1.37 shows the model track testbed. In order to simulate the increase in rail axial force due to temperature rise, he used electric currents to heat the rail.

A single rail was used to express the rigidity of the track panel by its bending rigidity, the sleeper was made to be 1/10 times the size of the real sleeper , and the ballast resistance was changed by changing the distance between the sleepers.

As shown in Fig. 1.38, Numata measured lateral ballast resistance by pulling all five sleepers at once and reading the load and displacement with a spring balance and a dial gauge to ensure measurement accuracy.

Since it was known from his theoretical analysis that initial track deviation significantly affected buckling, the laying of the model track was carried out with

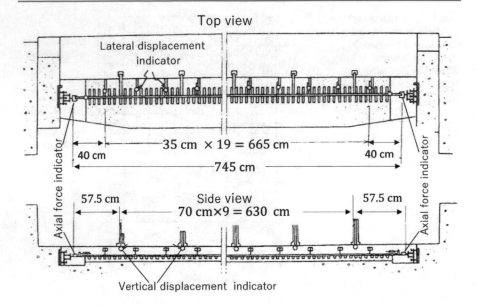

Fig. 1.37 Test track bed. *Source* Numata, M., "Buckling Strength of Long Welded Rail", Railway Technical Research Report, no. 721, RTRI, 1970, p. 64

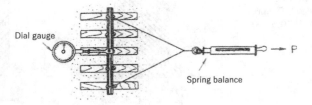

Fig. 1.38 Measurement of lateral ballast resistance. *Source* Numata, M., "Buckling Strength of Long Welded Rail", Railway Technical Research Report, no. 721, RTRI, 1970, p. 67

extreme caution. The rail was hung in the air by a pulley to ensure that there was no original strain, and then mounted on sleepers. The assembled track panel was lowered onto a 10-mm-thick sand testbed and then adjusted to achieve a horizontal deviation of less than ±0.1 mm and a vertical deviation of less than ±0.2 mm for the piano wire taut and parallel to the rail. Numata conducted such model tests 470 times, all of which required careful preparation and meticulous work (Photo 1.13).

The results of the model experiments were almost in agreement with the results of Numata's analysis. The next stage would be a demonstration on an actual track. Numata described about his work as follows [15]:

According to the theoretical studies and model experiments conducted so far, it is considered that continuous welding is possible without special reinforcement up to a curve

Photo 1.13 A scene from a
scale model experiment[16]

radius of about 600 m in terms of buckling strength, but it must be supported by an actual
track experiment. Since real track experiments require a very large-scale facility, this kind
of demonstration experiment is not enough even from a global perspective. Thus, we
decided to set up a particular test track at a yard and conduct actual large-scale experiments.
(Author's translation of the Japanese)

Figure 1.39 shows the test track with a length of 320 m and a radius of 600 m.
The buckling section was at the center of the track. Three steam locomotives were
located at both ends of the track, and 240 anti-creeping piles were driven into the
sections at both ends. In addition, 52 and 76 piles were driven in the sections to
prevent buckling at both sides of the buckling section. To heat the rails, pipes were
installed along the belly of the rail, and heated steam was fed to the pipe.

Tests began in June 1956 and were conducted eight times. On the morning of a
test, the rails were cut in the buckling section, rail clips were loosened to release
residual stresses, and the rails were re-welded. The rail temperature at that time was
the starting temperature of the test.

Photo 1.14 shows a test in progress, and Fig. 1.40 shows a comparison of the test
results and the calculated values.

This test validated Numata's analysis and allowed the CWR track to become
popular in Japan.

As described previously, it was not until the end of 1956 that the CWR track, an
essential technology for high-speed railways, became possible. This was two years
before the start of the Shinkansen project.

---

[16] Numata, M., "Buckling Strength of Long Welded Rail," *Railway Technical research report*, n
o.721, RTRI, 1970, p. 65.

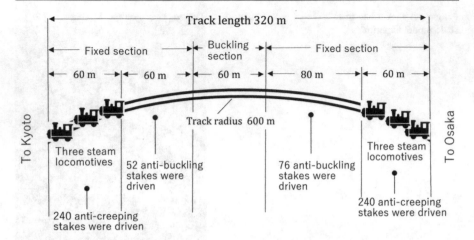

Fig. 1.39 Test track. *Source* Tachibana, F., Numata, M. et al., "Buckling Experiment of Curved Track", RTRI Research Materials, no. 7, RTRI, 1957, p. 8

Photo 1.14 Buckling test. *Source* Tachibana, F., Numata, M. et al., "Buckling Experiment of Curved Track", RTRI Research Materials, no. 7, RTRI, 1957, p. 13

1.3.2   Beginning of Track Dynamics

a. Measurement of Track Vibration and Stress

Tracks are structures that are used while repairing the deterioration caused by trains. Ballast tracks consist of rails, rail fastening devices, sleepers, and ballast. To design a track that suits the purpose, it is necessary to know what force each part of the track will be subjected to by trains, how each part will react, and how much damage each part will suffer.

Fig. 1.40 Comparison of calculated and measured values. *Source* Tachibana, F., Numata, M. et al., "Buckling Experiment of Curved Track", RTRI Research Materials, no. 7, RTRI, 1957, p. 15

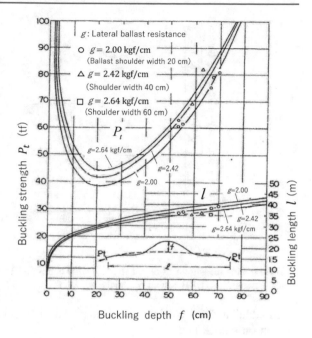

Buckling depth f (cm)

Research on so-called track dynamics began in the early 1950s by Sato,[17] who moved from the Navy to RTRI in 1946. In the preface of his research report, Sato states [16]:

> The purpose of this paper is to clarify the relationship between the various track deformations and train speed and to find a way to reduce the effect of speed. (Author's translation of the Japanese)

The measurement of displacement, stress, and vibration acceleration of various parts of tracks was made possible by Nakamura's strain gauge as described in Sect. 1.2.

Figure 1.41 shows the stress measured at the top and bottom of a rail, where the horizontal axis is speed and the vertical axis is stress. Since the measurement was made on December 2, 1950, it was probably the first such data derived in Japan.

Sato noted that the stress values did not change much as the speed increased. He also measured data at a rail joint , and again, the data was independent of the train speed. Sato said [16]:

> Contrary to expectations, the stresses do not increase with speed. The current method of calculating track stresses, which states that rail stresses increase by a factor of 1% at a vehicle speed of 1 km/h, is far different from what is actually measured. (Author's translation of the Japanese)

Sato's research benefited greatly from the advent of the strain gauge.

---

[17] Yutaka Sato joined the Imperial Japanese Navy in 1943, moved to RTRI in 1946 as a senior researcher, and later became head of the track laboratory of RTRI.

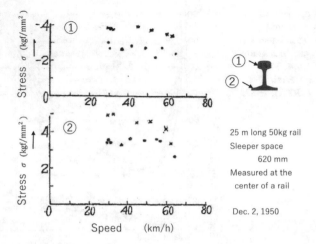

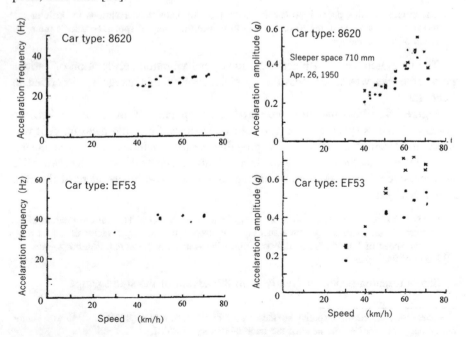

Fig. 1.41 First measurement of rail stress. *Source* Sato, Y., "The Influence of Train Speed upon Track Strength", RTRI Research Materials, no. 18, RTRI, 1952, p. 24

In addition, Sato obtained the data shown in Fig. 1.42 for the vibration acceleration on a track bed.

The left-hand figures show the acceleration frequency of the track bed, and the right-hand figures show the acceleration magnitude. It seems that the frequency does not depend on the train speed, but the amplitude increases with speed. On this point, Sato said [16]:

Fig. 1.42 Measurement of track bed acceleration. *Source* Sato, Y., "The Influence of Train Speed upon Track Strength", RTRI Research Materials, no. 18, RTRI, 1952, p. 25

If the vibrations are transmitted from cars, they should be proportional to the speed. But they are not actually, so the vibrations are considered the natural vibration of the track itself. However, since no equivalent vibration is found in the rails, it is assumed that the part below the sleepers mainly causes the vibrations. (Author's translation of the Japanese)

Sato derived the following from the measurements:

(i)   The amount of rail sinking by trains does not increase with speed.
(ii)   Bending stress of rails is also independent of train speed.
(iii) The vibration acceleration of the track bed is mainly in the range of 30 to 40 Hz, regardless of train speed, and seems to be the natural frequency of the part below sleepers.
(iv)   The acceleration magnitude of the track bed increases in proportion to train speed.

Regarding item (i) in the preceding list, Sato explained that rail sinking is not related to train speed because the propagation speed of rail deformation is about 1,000 km/h, much faster than train speed, and that speed effect would not appear in rail sinking below about 500 km/h.

b. Analysis of Track Vibration

What kind of calculation model can represent the aforementioned vibration characteristics? After examining several vibration models, Sato found that the model shown in Fig. 1.43 can represent a track. Since the rail sinking has no speed dependence, he thought, by simulating trainloads with a wheelset dropped from a certain height, it would be possible to know how track's mass and spring constants are related to each vibration.

However, in this model, the constants are distributed along the track, so the formula is complicated and difficult to calculate by hand. Computers were not available at that time. Therefore, Sato converted this model into an approximate lumped constant model, as shown in Fig. 1.44, and made it possible to use manual calculations.

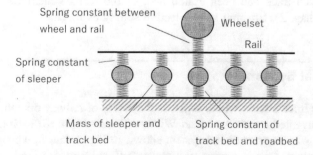

Fig. 1.43 Model for calculating track vibration characteristics (distributed constant model). *Source* Sato, Y., "Theoretical Consideration on the Relation between Structure of Railway Track and Vibration", RTRI Research Materials, no. 8, RTRI, 1956, p. 18

Fig. 1.44 Model for calculating track vibration characteristics (lumped constant model). *Source* Sato, Y., "Theoretical Consideration on the Relation between Structure of Railway Track and Vibration", RTRI Research Materials, no. 8, RTRI, 1956, p. 18

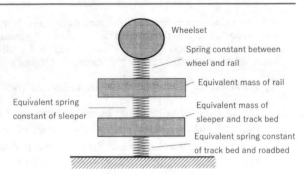

With this model, Sato calculated how the track vibration characteristics changed by changing the track's various parts and obtained the following results:

(i)   When the rail is made more massive, the displacement of axle, rail, and track bed decrease slightly, high-frequency component of the axle acceleration increases, and high-frequency portion of the rail acceleration decreases.

(ii) When the mass of the sleeper and the track bed is changed, the track bed's displacement does not change much, but the acceleration changes.

(iii) When the sleeper spring constant is increased, the displacement of each part decreases slightly, the intermediate frequency component of the axle acceleration increases, and the intermediate frequency component of the track bed acceleration increases considerably.

Based on these results, Sato concluded that the following are effective in reducing track destruction:

(iv) Making track beds thicker and heavier

(v)   Softening the spring between rails and sleepers with rubber pads or the like

(vi) Using crushed stone for the track bed

As a result of the research conducted through the mid-1950s, the dynamic characteristics of tracks had been elucidated and the basis for the rational design of the Shinkansen track had been established. In the RTRI's memorial lecture (described in Chap. 2), Hoshino gave a lecture stating as much.

1.4   Signal Systems

The fail-safe track circuit, which supports the safety of railway operation, is said to have been invented by an American William Robinson in 1872. Robinson's first track circuit was an open-circuit system shown in Fig. 1.45a, in which the electric light that signaled train detection was always off and turned on only when a train entered the track circuit.

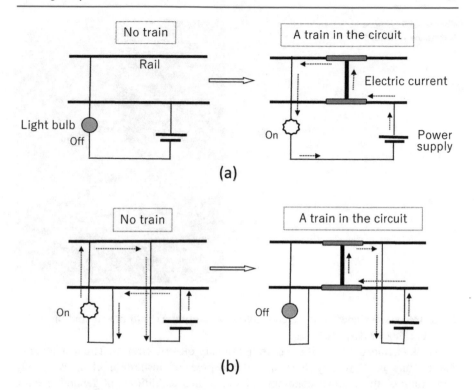

Fig. 1.45 Open and closed track circuit

However, with this system, if the equipment breaks, the conductor that conducts the signal current to the rails breaks, or a rail breaks, the light will not receive the signal current, even when a train enters the track circuit. This defect is corrected by a closed-circuit system shown in Fig. 1.45b, which lights up at all times and turns off when a train enters the track circuit. Thanks to Robinson's fail-safe signaling system using tracks, railways made great strides. After that, when railways were electrified, a large DC for power flowed through the rails, so DC track circuits could not be used, and AC track circuits of commercial frequency were used. The AC track circuit was introduced in Japan in 1913. Later, AC electrification of commercial frequencies with good cost effectiveness began, and AF (short for audio frequency) track circuits were developed. In recent years, digital track circuits incorporating communication technology developments have been created to further advance train control.

Army technical officer Hajime Kawanabe,[18] who later realized ATC (short for automatic train control) for the Shinkansen, moved to the Railway Technical Research Institute in December 1945 at the end of the war, as did Matsudaira and

---

[18] Hajime Kawanabe joined the Imperial Japanese Army in 1941, moved to RTRI in 1945 as a senior researcher, and later became the head of the signal laboratory at RTRI.

Miki. He was assigned to the electrical circuit laboratory and began studying track circuits (Photo 1.15).

Track circuits are very different from ordinary electric circuits. The insulation of the circuits is extremely low, and the degree of insulation changes greatly depending on the weather conditions and track bed conditions. In general, electric circuits are designed to prevent electric leakage, but track circuits are premised on considerable electric leakage. Railway signaling technology is a unique technology that has been developed to enable safe train operation with such electrically unstable track circuits.

Kawanabe, who was a communication engineer, began to clarify the electrical characteristics of this unique track circuit.

1.4.1 Simplification of Calculation of Track Circuit Constants

Since the track circuit is an electrical signal transmission line, its electrical constants (track circuit constants) consist of rail impedance and leakage admittance between both rails. Rail impedance is composed of rail resistance R and rail inductance L, and leakage admittance is composed of leakage conductance G and distributed capacitance C. Conductance, the reciprocal of resistance, represents the ease of current flow and is measured in mho.

These electrical constants cannot be directly measured because they are continuously distributed in rails, as shown in Fig. 1.46. However, it is known that the voltage V and current I at the track location x and time t are described by the following equation, called the telegraphic equation:

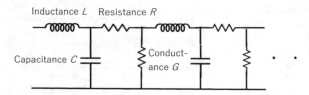

Inductance L Resistance R

Capacitance C

Conduct-
ance G

Fig. 1.46 Electrical constants distributed in the rails

$$C\frac{\partial V}{\partial t} + GV + \frac{\partial I}{\partial x} = 0$$

$$L\frac{\partial I}{\partial t} + RI + \frac{\partial V}{\partial x} = 0$$

Therefore, a method of calculating these constants by applying the voltage and current (including their phases) measured at the sending terminal with the receiving terminal open and short circuited to the telegraphic equation was used. Figure 1.47 shows the voltage and current at the sending terminal when the receiving terminal is open and shorted, V_o, I_o, V_s, I_s. However, this method was time consuming and laborious because it involved hyperbolic functions, which were very tricky to calculate manually. As stated repeatedly, computers were not available yet.

Therefore, Kawanabe first obtained the short-circuit impedance Z_s $(= \frac{V_s}{I_s})$ and the open-circuit impedance Z_o $(= \frac{V_o}{I_o})$ from the voltage and current at the sending terminal with the receiving terminal shorted or opened; he then created a nomogram, as shown in Fig. 1.48, by using these impedances and a newly introduced parameter P, which made it easy for anyone to calculate the track circuit constants.

In Fig. 1.48, the horizontal axis is the absolute value of the parameter P; and the vertical axis is its phase. One curve group is for the absolute value of $\frac{Z_s}{Z_o}$, and the

Fig. 1.47 Short circuit and open circuit of the track circuit

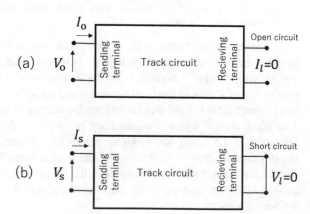

Fig. 1.48 Nomogram for reading electrical constants of track circuit. *Source* Kawanabe, H., "The Calculation Method of AC Track Circuit Constant", Signal Engineering of Japan, no. 3, The Signal Association of Japan, 1951, p. 67

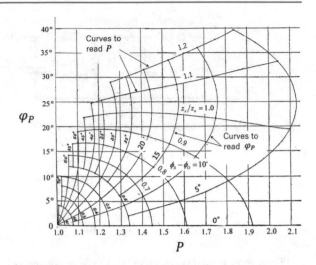

other group is for its phase. Calculate the absolute value and phase of $\frac{Z_s}{Z_0}$, apply it to the chart, read P from the intersection of both curves, and use this P to obtain $Z = PZ_s$, $Y = P/Z_0$. Since $Z = R + j\omega L$ and $Y = G + j\omega C$, R, L, G, and C can be obtained.

This method's effectiveness was substantial, and it significantly contributed to improving the efficiency of track circuit research.

1.4.2 Development of AF Track Circuit

In 1951, five years after Kawanabe came to RTRI, he began to focus on AF track circuits. He thought [17]:

> What would happen if I dare passed a high-frequency current through two rails to form a track circuit? Of course, with the frequency increased, the current will be attenuated more, but if vacuum tubes or transistors are used, amplification will counteract the attenuation. Using rails as carrier transmission lines could revolutionize train control. (Author's translation of the Japanese)

Before Robinson invented the track circuit, train drivers watched trains' forward sections and drove to avoid collisions. However, this method of driving is dangerous when visibility is poor. With the invention of the track circuit, the presence or absence of a train in the front section was displayed on the signal light beside the track installed far behind that section, so the degree of safety was greatly improved. The track circuit was a revolutionary advance in the safe driving of railways. However, if the driver inadvertently passes by without looking at the signal light, the device is ineffective. Kawanabe's idea was to make it possible to create a device that would always display the signal in the driver's seat, thus eradicating accidents caused by driver inattention. In other words, it was the second revolution after the Robinsob's track circuit.

Kawanabe learned that in 1944, before he came to the RTRI, work on rail-based communication had already been done by Hirokawa and others,[19] and he began to measure the electrical constants of track circuits in the AF band with reference to Hirokawa's work.

In 1953, he posited that a 3-km track circuit could be made possible using a frequency of 1 kHz. As for the frequency to be applied, he concluded that about 1 kHz was an appropriate choice in transmission loss and demodulation stability.

Kawanabe wanted to realize an AF-based track circuit device with transistors. The point-contact transistor had been invented by W.H. Brattain and J. Bardeen of Bell Telephone Laboratories in 1947, followed by the junction type by W. Shockley in 1948. They were awarded the Nobel Prize in Physics in 1956. Transistors work at low voltage, consume little power, are resistant to vibration, and have long life. Kawanabe thought that the transistor was the perfect device for railway signaling equipment.

The transistor, a historic invention, which was based on solid-state physics, later developed into computers and other sophisticated electronic devices, giving birth to the advanced technological society we know today.

In Japan, in 1953, the development of AC electrification with commercial frequencies began. In the AC electrification, the conventional AC track circuit with commercial frequencies could not be used. So, the AF track circuit was used, but one big problem was how to design it to be fail-safe.

Traditionally, the electrical components used in railway signaling equipment had been stable ones such as transformers, reactors (all made of iron and copper), resistors (insulators and carbon or metal coatings), and relays that operate by gravity if there is no driving power. These were used with a fail-safe concept (i.e., configuring the system so that it stops the train in the event of a failure). The advent of the AF track circuit with electronic circuits revolutionized the thinking in this extremely cautious field of railway signaling technology. Electronic circuits, such as amplifiers, do not always stop the train when they are broken. A major challenge was how to incorporate electronic circuits into railway signals in a fail-safe manner.

Another challenge was to select a modulation method that could protect signal currents flowing in track circuits from the noise of driving currents, which were hundreds of times greater than signal currents. In this regard, Kawanabe compared the amplitude modulation and pulse modulation on a DC electrified line and chose the amplitude modulation. (This is the modulation method used in radio broadcasting.)

Regarding the fail-safe design, Itakura,[20] who worked with Kawanabe on developing the AF track circuit, stated [18]

---

[19] Genji Hirokawa, Susumu Ukai, Kensaku Takahashi, "Electrical Characteristics of Track Circuits at Audio Frequency Domain" (in Japanese), *RTRI Research Materials*, n o. 1, RTRI, 1944.

[20] Eiji Itakura joined JNR in 1955. After serving as a senior researcher at RTRI, he became the head of the signal laboratory of RTRI.

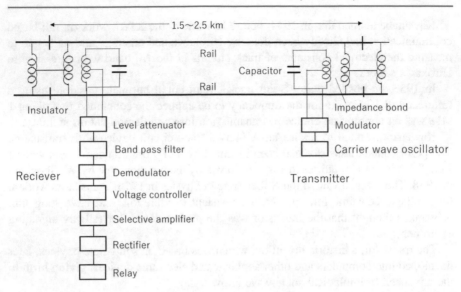

Fig. 1.49 First signaling device configuration using AF track circuit. *Source* Kilocycles Track Circuit Research Report, The Signal Association of Japan, 1957, p. 43

It can be said that efforts to fail-safely design electronic circuits used for important parts such as train detection in adverse conditions on the ground were big challenge. We hesitated to use transistors, which were in their infancy, and used proven vacuum tubes. Even those were limited to the high-reliability tube for communication, carefully selecting the ones with good characteristics.

… Measures to prevent track relays from malfunctioning due to noise wave when there is no signal wave were carefully examined from the viewpoint of fail-safeness, and many safety technologies were incorporated. (Author's translation of the Japanese)

Having solved the fail-safe problem, Kawanabe designed a signaling device for the first AC-electrified Hokuriku line in 1957, using the AF track circuit.

Figure 1.49 shows the equipment configuration of the signal system using the AF track circuit. For the carrier wave, 720 Hz was used for the inbound line, 960 Hz for the outbound line, and 20 Hz for the signal wave.

Figure 1.50 shows the current waveforms flowing in track circuits. There are three frequencies: carrier wave, carrier–signal wave, and carrier + signal wave. The onboard equipment picks these up from the rails and demodulates them to reproduce the original signal wave.

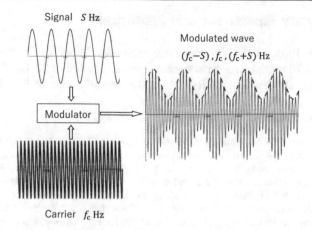

Fig. 1.50 Waveforms flowing in the AF track circuit

1.4.3   Development of Cab Warning Device

The fail-safe AF track circuit was not yet completed, but Kawanabe tested a first cab warning device in 1954. The test attempted to send three signals of proceed, caution, and stop onto the vehicle; the device used vacuum tubes for the ground device and transistors for the onboard device. The onboard device emitted an alarm sound when the train approached a stop signal to alert the driver but did not automatically apply the brakes. Specifically, a signal current of 1,430 Hz was superimposed on the DC track circuit, periodic modulation was performed 300 times per minute for the proceed signal, no modulation (carrier wave only) was for the caution signal, and no carrier wave was for the stop signal. The onboard device picked up these waves and amplified them, restored the signals, and displayed them to the driver.

The first transistor was not released in Japan until February 1954, so the aforementioned onboard device used free sample transistors from a manufacturer. (Incidentally, transistor radios were first released by Sony in 1955). This device was improved and put into practical use as the cab warning device on the Tokaido Line in 1960. Kawanabe used 1,300 Hz for the carrier wave in this device, 20 Hz for the proceed signal, 35 Hz for the caution signal, and no signal (carrier wave only) for the stop signal.

Kawanabe's purpose was to revolutionize train control. As mentioned previously, through the practical application of the AF track circuit for AC electrification of the Hokuriku Line and the development of the cab warning device using the AF track circuit, he was one step closer to his goal of automatic train control by the end of 1956.

As will be described in Chap. 2, Director Takeshi Shinohara was appointed to RTRI in early 1957, and the idea of the Shinkansen emerged.

1.5   Catenary Equipment and Pantographs

On March 28, 1955, an electric locomotive of Société Nationale des Chemins de fer Français (SNCF) pulled three passenger coaches to a speed of 326 km/h, and on the following day, another locomotive achieved a record speed of 331 km/h (Photo 1.16).

Murray Hughes describes the events as follows [19]:

Two locomotives, the four-axle BB9004 and the six-axle CC7107, had been chosen for a daring attempt to test the limits of railway technology. Each was matched with a test train consisting of three third-class carriages that had been fitted with some rudimentary aerodynamic improvements such as a rounded 'tail' on the last car. On 26 March the BB locomotive hauled its three test cars along the long, straight stretch of track at an unprecedented 276 km/h. Two days later CC7107 crushed this astonishing achievement with a dynamic sprint that smashed through 300 km/h for the first time in railway history. It was truly spectacular, with linesiders watching in horror as the locomotive's pantograph melted, setting fire to nearby pine trees. And on 29 March the BB locomotive outpaced its sister with another breakneck dash—again with a dangerous firework display and ballast fragments blasted into the air by the slipstream. SNCF proudly announced that both locomotives had reached the incredible speed of 331 km/h. Many years later this was revealed not to be true, but at the time SNCF was anxious not to favour one locomotive manufacturer over another. In fact, only the BB had reached the maximum speed, the CC having attained a no less creditable 326 km/h. For many years SNCF concealed what had really happened during the final record run. Observant spectators who witnessed the event and saw the state of the track after the final passage of the BB9004 were instructed to keep mum. Eventually, the French magazine *La Vie du Rail* published a photograph showing that BB9004 and its train had seriously bent the rails. This had been caused by a bogie thrashing furiously from side to side as the train hurtled along the track, a phenomenon known as 'hunting'. The wheels had exerted so much lateral force that the rails were moved out of alignment. Miraculously, the train did not derail, and the story of high-speed railways moved on

This description tells us how close the test run came to a derailment (Photo 1.17).

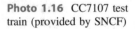

Photo 1.16 CC7107 test train (provided by SNCF)

Photo 1.17 Track
irregularity after the test run
(provided by La Vie du Rail)

Masaru Iwase,[21] who was in charge of developing the sliding strips of Shinkansen pantographs, wrote in his book about the power collection of this run as follows [20]

> According to records, of the electric locomotive's two pantographs, the front pantograph was lowered, and only the rear was raised to start. Just before the pantograph was destroyed by severe arcing (as expected by engineers), the front pantograph was raised and the rear was lowered to continue accelerating. It was not surprising because currents were about 4,500 amps at 1,500 V DC, and contact loss ratios were about 25%. (Author's translation of the Japanese)

These descriptions indicate that the technical barriers to speeding up trains were railcars' hunting motion and the difficulty in power collection, and that dangerous hunting did not occur with CC locomotive with three-axle bogies but with BB locomotive with two-axle bogies. Since the research progress of the hunting motion has already been described, the study of the current collection system, which was another barrier, will be traced below.

Tracks and vehicles are highly independent, so there is no problem even if axle distances (wheelbases) of bogies, vehicle length, or the number of connected vehicles changes. However, in power collection systems, the degree of mutual influence between pantographs and catenary equipment is large; that is , pantograph intervals and the number of pantographs affect power collection quality. When train

---

[21] Masaru Iwase joined the Ministry of Transport in 1945. After served as a senior researcher at RTRI, he later became the head of the power collection laboratory of RTRI and a professor at Nippon Institute of Technology.

speed is increased, even in a single pantograph, the vibrations of overhead wires reflected at support points may give an adverse effect on pantograph motion.

The fundamental problem of such power collection systems is not a problem when trains are at low speed. It suddenly becomes apparent at high speed, and the contact loss between pantograph and contact wire increases, causing sparks and sometimes leading to the melting of pantograph strips. The broken pantographs by severe sparks at the time of the aforementioned speed record in France seem to show this situation clearly.

1.5.1 Establishment of a Research Committee on Power Collection

Research on power collection systems began in earnest in 1952 when the Railway Electrification Association established a committee for power collection research. Although the Tokaido Line was still undergoing electrification work, there was a plan to improve train speed when the work was completed, and there was concern that the power collection system would become a bottleneck.

The research group started a worldwide literature review on power collection and found that there was no research on the dynamics of catenary–pantograph systems.

The difficulty of research in this field is the high voltage on the overhead equipment and pantographs, making existing measuring devices almost useless. Therefore, the study group started with the development of measuring instruments. The left side of Fig. 1.51 shows a newly developed vertical movement sensor of contact wires, and the right side shows a oscillogram obtained by this sensor.

The sensor weighs only 20 g, so it does not interfere with the movement of contact wires. In addition, an acceleration sensor for pantographs and a contact loss sensor were developed for the high-speed test scheduled for the following year. For contact loss detection, a minute current of about 10 mA was collected from the pantograph dedicated to measurement, and the current interruption was defined as

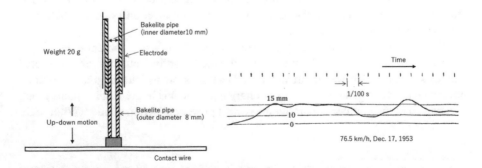

Fig. 1.51 Vertical motion sensor for the contact wire and measurement results. *Source* Research on Power Collection (1), Railway Electrification Association of Japan, 1955, p. 45

the contact loss (because for a current of this level, no arc is generated by contact loss).

Measuring pantograph acceleration was dangerous hotline work because there were no wireless telemeters at that time. A measurement cable charged to 1,500 V was pulled into a car and connected to measuring instruments placed on an insulated floor. Engineers performed measurement work on this floor (they sometimes had to put on hotline protective wear). Even the slightest carelessness could lead to an electric shock. Surprisingly, in the Shinkansen test section (see Fig. 4.12), measurement was performed by 25,000-V hotline work, which would be unacceptable today. The situation improved in 1968 when wireless telemeters became available for such dangerous work.

1.5.2   120-km/h High-Speed Test

A report of the Current Collection Study Committee states [21]:

> The maximum speed limit on straight sections is currently 95 km/h for passenger trains, which was decided in 1924. Given the restoration and maintenance conditions of rolling stock and tracks after the war, speeds exceeding 95 km/h would be possible, at least on first-class tracks.
>
> ... If the speed limit can be increased to 120 km/h in straight sections and can be increased considerably in curved sections and turnouts, the travel hour between Tokyo and Osaka can be reduced to five and a half or six hours, allowing one round trip within a day with the same train. (Author's translation of the Japanese)

As the Tokaido Line's electrification work progressed steadily, the behavior of vehicles, tracks, and power collection systems was measured in March 1954 to gather data for raising the maximum operating speed to 120 km/h. Figure 1.52 shows the test train formation.

The power collection system consisted of simple catenary equipment with one messenger wire and one contact wire (see Fig. 1.56) and PS13-type pantographs. Figure 1.53 shows a contact loss measurement at 120 km/h.

Large contact loss reaching about 1/4 of the span length of 50 m occurred at each support point. Therefore, it is supposed that the train ran while emitting large arcs, and pantograph strips suffered considerable melt damage. Today, even in a test

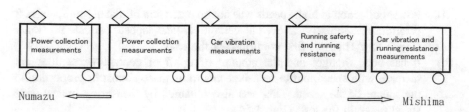

Fig. 1.52 Test train formation. *Source* Arimoto, H., From Kodama to Hikari, Osaka Dengyo Co. Ltd., 1976, p. 16

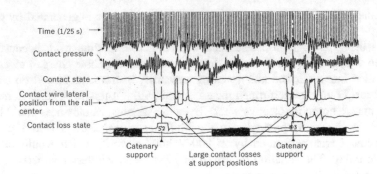

Fig. 1.53 Large contact loss of span cycles (120 km/h). *Source* Test Report on High-Speed Power Collection, Railway Electrification Association of Japan, 1954, p. 11

Fig. 1.54 Speed and contact wire uplift. *Source* Test Report on High-Speed Power Collection, Railway Electrification Association of Japan, 1954, p. 17

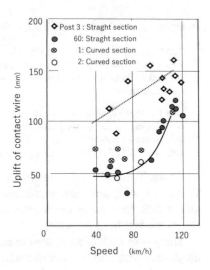

situation, such running would not be allowed. Figure 1.54 shows the maximum uplift[22] of the contact wire with respect to speed, and Fig. 1.55 shows the relationship between speed and contact loss.

The test report states that:

(i) Power collection at high speeds was not an extension of that at low speeds . It deteriorated abruptly above a certain critical speed, about 90 km/h (Fig. 1.55). This was first confirmed at the tests.

(ii) It has been thought that the amount of uplift of contact wires does not increase with speed, but in fact, it seems to increase almost in proportion to the square of the speed. This was also obtained by measuring over a wide speed range in the tests (Fig. 1.54).

---

[22] The "uplift" is the amount of displacement of contact wire pushed up by pantographs.

Fig. 1.55 Speed and contact loss. *Source* Test Report on High-Speed Power Collection, Railway Electrification Association of Japan, 1954, p. 11

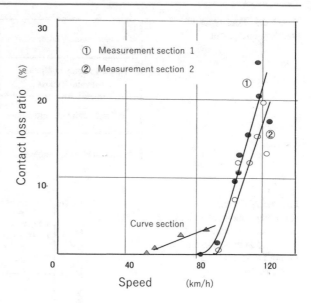

(iii)   The large contact losses that appeared above the critical speed are considered to be caused by pantograph's inability to follow the difference in displacement of the contact wire between supports. Since the pantograph used was PS13 with poor characteristics, the critical speed may increase with a different model with better characteristics. If pantograph's push-up force is temporarily increased, the power-collecting performance should improve, although it might cause other adverse effects.

Although the measurements on catenary equipment and pantographs had been made in 1948, the data obtained at this test with the newly developed measuring instruments were a starting point for developing high-speed power collection systems. With these findings, the next step was to test the different types of catenary equipment.

1.5.3   Comparative Test of Four Types of Catenary Equipment

From August 1955 to February 1956, in order to decide the policy needed for improving the overhead equipment of the Tokaido Line in preparation for a new 120 km/h train composed of electric railcars, a comparative test of four types of catenary equipment shown in Fig. 1.56 was conducted. The pantographs were two PS13 types, and the speeds were 60, 90, and 120 km/h.

Figure 1.57 shows the contact loss tendency of each catenary equipment. The test results were summarized as follows:

Fig. 1.56 Four types of
catenary equipment

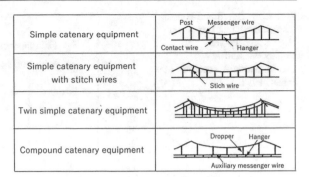

Fig. 1.57 Catenary
equipment and contact loss
ratio. *Source* Arimoto, H.,
From Kodama to Hikari,
Osaka Dengyo Co. Ltd.,
1976, p. 16

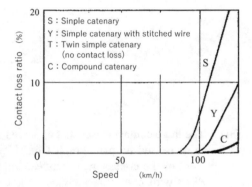

(i)   Simple catenary equipment cannot be used at speeds above 100 km/h due to
       the rapid increase in contact loss and uplift.
(ii)   Simple catenary equipment with stitch wires can increase the speed by 15 or
       20 km/h as compared to the simple catenary type.
(iii) Compound catenary equipment and the twin simple catenary equipment did
       not cause any conspicuous problems even at 120 km/h. In particular, the twin
       simple catenary type did not show any abnormalities.

However, it was still not known what led to these results.

The catenary equipment of the Tokaido Line was upgraded in light of the test
results. In November 1958, the limited express train of electric railcars named
Kodama (which means echo) began operating at 110 km/h, reducing the travel time
between Tokyo and Osaka from 7 hours and 30 minutes to 6 hours and 50 minutes.

1.5.4 New Approach to Power Collection System Dynamics

Having found that the simple catenary equipment causes large contact loss at a
speed of over 100 km/h, the study group started to investigate its cause and to take
countermeasures.

As stated previously, the results of the speed improvement tests so far were understood as follows:

(i) The reason for the large contact loss above the critical speed is that the pantograph cannot follow the catenary equipment's deformation between the support points.
(ii) Since the pantograph used for the test was PS13 type with poor characteristics, the critical speed may be increased by using the one with better characteristics. In addition, if the push-up force is increased temporarily, the contact loss will decrease, although there will be other harmful effects.

These statements reveal the thinking at that time regarding the cause of contact loss. That is, a pantograph travels while pushing up contact wire, but because the amount of wire's uplift is different between support points and the center of spans, the pantograph runs in up-and-down motion. The higher the speed, the faster the up-and-down movement and the greater the pantograph's inertia force. If the inertia force becomes larger than the push-up force, the pantograph will separate from the contact wire, so the stronger the pantograph's push-up force, the less likely it separates.

However, Sumiji Fujii,[23] a member of the study group, believed that the conventional understanding cannot explain the phenomenon and submitted a paper to the study meeting in September 1955, in which he considered catenary equipment and a pantograph as one vibration system and clarified its essential vibration characteristics; as shown in Fig. 1.58b, he represented the catenary equipment by a tensioned string with no mass, and the pantograph by the mass M and the push-up force P_0. Then, as shown in Fig. 1.58c, he further simplified the catenary equipment by a spring $k(x)$ whose strength changes depending on the location.

Fujii defined the variation of the catenary system's spring constant within one span as a sinusoidal wave, which is

$$k(x) = K\left(1 - \varepsilon \cdot \cos\frac{2\pi}{S}x\right) \tag{1.5}$$

where K is the average value of spring constants, ε is the inequality ratio of spring constants, and S is the span length.

An approximate solution of the equation of motion for the model shown in Fig. 1.58c shows that the system has a resonant velocity V_c and a velocity V_r where contact loss starts to appear.

---

[23] Sumini Fujii was an assistant professor in the Department of Mechanical Engineering at the University of Tokyo at the time; he later became a professor at the university, then dean of the Faculty of Engineering, president of Japan Society of Mechanical Engineers, president of Japan Society of Robotics, and president of Toyama Prefectural University.

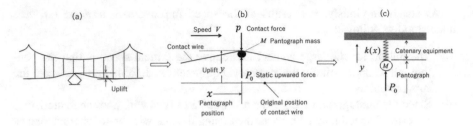

Fig. 1.58 Modeling of catenary equipment and pantograph

$$V_c = \sqrt{1 - \frac{1}{2}\varepsilon^2} \cdot \frac{S}{2\pi}\sqrt{\frac{K}{M}}, \; V_r = \sqrt{\frac{1 - \frac{\varepsilon^2}{2}}{1 + \varepsilon}} \cdot \frac{S}{2\pi}\sqrt{\frac{K}{M}} \tag{1.6}$$

Fujii separately clarified the effect of adding damping to the pantograph. Applying a damper is out of the question in the conventional view since it reduces the pantograph's movability. However, from the point of view that the catenary equipment and pantograph are integrated vibration system's components, it is only natural that damping should be added to the system to reduce resonance. The results of Fujii's analysis revealed the following:

(1) In order to increase V_c and V_r, the following are effective:

    (i)   Increase the average spring constant K of catenary equipment.
    (ii)   Decrease the inequality ratio ε of spring constants.
    (iii) Decrease the pantograph mass M.

(2) Appropriate damping of pantographs will suppress the resonance and prevent the contact loss.

Traditionally, improved pantographs and increased push-up force were thought to be effective in reducing contact loss , as described, but the role of catenary equipment was not considered. Fujii's theory was the first to clarify the role of both the catenary equipment and pantograph by dynamically analyzing them as one integrated vibration system, thereby opening the door to high-speed power collection systems.

It was a year and a half before the RTRI memorial lecture, which will be described in Chap. 2. The Shinkansen's power collection system, which will be described in Sect. 4.2, was developed based on Fujii's theory.

References

1. Tadashi Matsudaira, "From Zerofighter to Shinkansen (in Japanese)," Journal of the Japan Society of Mechanical Engineers, no. 667, 1974, p. 32.
2. Tadashi Matsudaira, "A Retrospective of Research and Development on the Tokaido Shinkansen (in Japanese)," Journal of the Japan Society of Mechanical Engineers, no. 646, 1972, p. 102.
3. Tadashi Matsudaira, "Natural Frequencies of Coarches and Electric Railcars (in Japanese)," RTRI Research Materials, no. 2, Railway Technical Research Institute (RTRI), 1949, p. 3.
4. Tadashi Matsudaira, Masaharu Kunieda, Keiji Yokose, "Theory of Forced Vertical Vibration of Bogie Cars, with Special Reference to Optimum Values of Stiffness of Spring and Friction (in Japanese)", RTRI Research Materials, no. 8, RTRI, 1951, p. 9.
5. Tadashi Matsudaira, "Basic Theoretical Considerations on Hunting Motion of 2-Axle Bogie (in Japanese)," Paper no.63 at the third study meeting, 1947.
6. Masaharu Kunieda, "Measurement of Train Vibration (in Japanese)," Paper no. 101 at the fourth study meeting, 1948.
7. Tadashi Matsudaira, "Shimmy of Axle with a Pair of Wheels (in Japanese)," RTRI Research Materials, no. 19, RTRI, 1952, p. 25.
8. Tadashi Matsudaira, Taiji Aizawa, Yoshio Mukaide, Keiji Yokose, "Train Speed-up through the Improvement of Spring Suspension System of 2-axle Freight Cars (in Japanese)", RTRI Research Materials, no. 18, RTRI, 1953, p. 5.
9. Tadanao Miki, "Reminiscence of Odakyu Type 3000 SE Car Design (in Japanese)", Japan Railfan Magazine, no. 375, Koyusha, 1992, p. 92.
10. Tadashi Matsudaira, "Memories of 24 Years in the Post-War RTRI (in Japanese)", 10 -Year History: 70th Anniversary, RTRI, 1977, p. 291.
11. Shuuzo Nkae, "Exploring the Roots of the Strain Gauge (in Japanese)", Railway Research Reviw, no.1, RTRI, 2009, p. 1.
12. Kazuo Nakamura et al., "The Dynamic Problems of the Wheel Axles of Railway Vehicles (Report 1) (in Japanese)", RTRI Research Materials, no. 11, RTRI, 1956, p. 8.
13 Tadanao Miki, "A Concept for the Super Express (4.5 Hours between Tokyo and Osaka) (in Japanese)", The Engineering Journal for Transportation, no. 89, Kotsu Kyuoryokukai, 1954, p. 2.
14. Yoichi Hoshino, "Thermal Expansion of Very Long Rails (in Japanese)", RTRI Research Materials, no. 2, RTRI, 1951, p. 16.
15. Fumikatsu Tachibana, Masahiko Tanaka, Hdeaki Suzuki, Minoru Numata, "Buckling Experiment of Curved Track (in Japanese)," RTRI Research Materials, no. 7, RTRI, 1957, p. 4.
16. Yutaka Sato, "The Influence of Train Speed upon Track Strength (in Japanese)", RTRI Research Materials, no. 18, RTRI, 1952, p. 1.
17. Hajime Kawanabe, "Aiming for Innovation in Signal Technology (in Japanese)", Signal Engineering of Japan, no. 11, The Signal Association of Japan, 1958, p. 25.
18. Eiji Itakura, "The Development of the AF Track Circuit (in Japanese)", History of the Development of Railway Signals, The Signal Association of Japan, 1980, pp. 282–283.
19. Murray Hughes, The Second Age of Rail - A History of High-Speed Trains, 2nd edition, The History Press, 2020, pp. 19–20.
20. Masaru Iwase, À la Carte Power Collection Technology (in Japanese), Kenyusha Foundation, 1998, p. 114.
21. Test Report on Power Collection of High Speed Trains (in Japanese), Railway Electrification Association of Japan, 1954, p. 1.

RTRI's Commemorative Lecture on the Possibility of High-Speed Railways

2

In 1957, RTRI celebrated the fiftieth anniversary of its founding. One of the commemorative events was a lecture that played a major role in realizing the Shinkansen. In his memoirs, Matsudaira said the following about the lecture [1]:

> In April 1957, the Railway Technical Research Institute celebrated its fiftieth anniversary and held a lecture entitled "The Possibility of a Three-Hour Tokyo-Osaka Service" on May 30 as one of the commemorative events. This event was the idea of Takeshi Shinohara,[1] the director of the research institute at that time, and it was an attempt to solve the problems that the Tokaido Line, which is the main artery of Japan, was having: demand for its services outpaced its transportation capacity at the time. In line with his idea of building an ultra-high-speed railway that is not bound by traditional formats, he tried to widely inform the public of its technical potential. (Author's translation of the Japanese)

Shinohara was 50 years old when he was appointed the director of the RTRI in January 1957. It was the first time he had ever worked for RTRI (Photo 2.1).

As part of his mission to take control of all aspects of the institute's operation, Shinohara received reports from the general affairs and accounting department, as well as from the heads of the various research sections, and he learned of the research results accumulated by the engineers who had been transferred from the military after the war.

In his book, Shinohara wrote the following [2]:

> Tadanao Miki was the chief design engineer for the Navy's Galaxy bomber during the war, and Tadashi Matsudaira was the man who conducted wind-tunnel tests on the flutter of Zero fighters to prevent vibration and disintegration of aircrafts in mid-air. Miki and Matsudaira were aircraft engineers, but after the war, Japan was banned from aircraft manufacturing, so they engaged in railway research. ... my alma mater later awarded me a doctorate in civil engineering, but I was a complete novice when it came to train design.

---

[1] Takeshi Shinohara joined the Ministry of Railway in 1930, later became Chairman of the Japan Society of Civil Engineers, and later became President of the Japan Railway Construction Public Corporation.

© The Author(s), under exclusive license to Springer Nature Singapore Pte Ltd. 2022
T. Shimomae, *Birth of the Shinkansen*,
https://doi.org/10.1007/978-981-16-6538-7_2

Photo 2.1 Takeshi
Shinohara (Provided by
RTRI)

However, I think I was able to listen to and understand what the engineers said and put it together in one concept. (Author's translation of the Japanese)

Whether it is technically possible to double the speed of conventional railways to 200 km/h when their current maximum speed is less than 100 km/h cannot be determined without solid insight into the wide range of technologies that comprise railways, including tracks, rolling stock, signal safety, and power supply. The researchers in each field have the best understanding of each technology's limitation and future possibility, so Shinohara summarized their opinions in preparation for the aforementioned lecture in honor of the institute's founding.

Shinohara wrote [2]:

At the meeting, it was decided to hold a research lecture for the general public, and to unify the research on high-speed rail transportation, which was being conducted separately at the time, into an easy-to-understand title. However, "Research on High-Speed Railway Transportation" was too abstract a subject for the lecture. In order to make the meaning of research topics clearer, we decided to entitle it "The Possibility of a Three-Hour Tokyo-Osaka Service." (Author's translation of the Japanese)

On the day of the lecture, it was raining heavily, and those involved were concerned that the inclement weather might deter potential attendees. However, the audience exceeded the event capacity, and organizers had to refuse admission when full capacity was reached (Photo 2.2).

The lecture began with a screening of a 16 mm color film, *In Search of New Electrification*, followed by Shinohara's welcome greeting, four lecturers' presentations (each of whom presented slides to accompany the lectures), and, finally, a documentary film on the SNCF's speed record of 331 km/h in 1955. The documentary film and the day's events ended at 4:30 p.m.

Photo 2.2 Commemorative lecture. (Provided by RTRI)

In his opening remarks, Shinohara said the following [3]:

As you all know, the transportation capacity of the Tokaido Line is very tight. So, we must decide whether to add the same line parallel to the current line or to build a new standard gauge line. Forward thinking, with our eyes on the future, indicates that we should build a standard gauge line. The content we are going to talk about here is the case of building a new track whose gauge is 1.435 m, since the current narrow gauge is unstable at high speeds. (Author's translation of the Japanese)

There had been a long history of debate in Japan about whether to use standard or narrow gauge tracks. The so-called bullet train project that would connect Japan with the Asian continent adopted the standard gauge. The bullet train was planned to connect Tokyo to Shimonoseki, a distance of about 1,000 km, in nine hours at a maximum speed of 200 km/h (Fig. 2.1). Construction of this project began in 1940, but was terminated in 1944 due to the worsening war situation.

In May 1956, 11 years after the end of the war, JNR established the Tokaido Line Reinforcement Study Committee. The committee held repeated discussions but was adjourned without reaching any definitive plan of action. But the RTRI's commemorative lecture held four months later had a great influence on the subsequent development (see Chap. 3).

Following Shinohara's speech, lectures were given on vehicle structure, running safety and ride quality, track structure, and signaling safety. The following sections summarize the presentations.

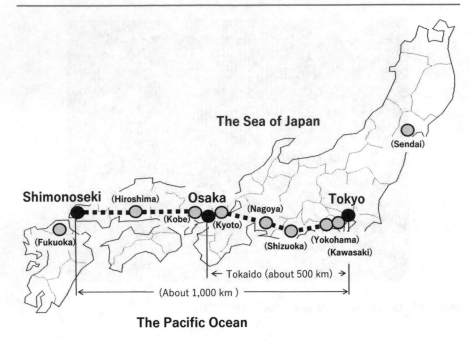

Fig. 2.1 Tokyo–Osaka–Shimonoseki

2.1 Vehicle Structures (Tadanao Miki)

The high-energy efficiency of railways is an advantage not found in other forms of transportation; therefore, as long as railways keep up with the demands of the times, they should not diminish as a medium-distance mode of transportation, although they will have to give up long distances to airplanes.

(1) **Speed Necessary to Connect Tokyo and Osaka in Three Hours**

    (i)   The distance between Tokyo and Osaka is about 400 km in a straight line or 559 km on the current railway line. It is estimated to be 450–500 km if major cities are connected in the shortest possible distance.

   (ii)   The ratio of scheduled speed[2] to maximum speed is 0.78 for the present limited express and 0.75–0.85 for a distance of 300–600 km in Europe and America. Therefore, if new line's curve radiuses are increased, the number of stops is reduced, and the number of places that require speed

---

[2] Scheduled speed is calculated by dividing the travel distance by the travel time, including stops along the way.

limits is reduced, the ratio can be 0.8. Based on these figures, to connect Tokyo and Osaka in three hours, the scheduled speed target should be 170 km/h, with a maximum speed of 250 km/h.

(2) Gauge

Massive horsepower is needed to reach 250 km/h, so to mount large output motors on bogies, the standard gauge or broader is required. However, if the gauge is wider than necessary, cars will be heavy.

(3) Shape of Vehicle

At high speeds, aerodynamic drag is proportional to the cube of the speed. Looking at France's 331-km/h-train speed-to-horsepower curve, at 300 km/h, 90% of the power is spent on air resistance. This means that it is essential to make the train streamlined at high speed. Wind-tunnel tests have shown that air resistance can be reduced to one-third of that of conventional trains if the train is close to a perfect streamline, the car height is lowered, the front bogie is covered, the roof vents are removed, and the couplers are covered with a hood to smooth the space between cars (Photo 2.3).

Although the impact of wind pressure when passing oncoming trains or entering a tunnels can be reduced by making the train streamlined, there are still research issues to be addressed, such as the shape of the tunnel entrance and track spacing.

(4) Downsizing and Weight Reduction of Vehicle

It takes a lot of energy to speed up a train, and this energy must be absorbed when stopping the train. Since this energy is proportional to the train mass, it is essential to lighten high-speed railway's vehicles. As an added benefit, light cars help to reduce the construction and maintenance costs of the track.

The weight reduction of railcars began around 1952. The weight of modern railcars is between 60 and 70% of that of conventional ones. It is possible to decrease the weight of cars even further by using lightweight alloy that is one-third lighter

Photo 2.3 Vehicle models used in the wind-tunnel test. (Provided by RTRI)

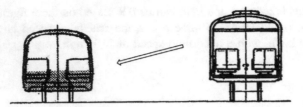

Fig. 2.2 Conventional vehicle and new vehicle with a low center of gravity. *Source* Studies of the Tokaido Shinkansen, vol. 1, RTRI, 1960, p. 9

than steel, similar to what is used for airplane construction. It is also possible to use synthetic resins, which have recently made remarkable progress. However, the lighter the vehicle, the lower its height and center of gravity must be, as shown in Fig. 2.2, to ensure stability and to reduce the risk of overturning due to crosswinds.

At high speeds, windows cannot be opened, so they must be fixed. Making the windows fixed would be structurally advantageous because of the better coupling between the top and bottom, but the train should have fully automatic air conditioning.

(5) Power System

Power systems can be of two types: a power-concentrated locomotive type and a power-distributed train type. The power-distributed system is helpful in reducing the loads on tracks and improving the efficiency of acceleration and braking.

Even a streamlined light train requires 3,000 to 4,000 horsepower to achieve 200 km/h. The necessary horsepower is possible if all bogies are equipped with motors. The center of gravity can be lowered by mounting equipment under the floor of each vehicle.

High-speed tests in France show that power collection is a major challenge. Research must be done on pantographs that have low aerodynamic drag and low contact loss.

(6) Braking Method

Trains running at high speed must, of course, stop. In an emergency, they must be stopped in the shortest possible time. Therefore, braking is one of the most important research issues for high-speed rails. For high-speed trains, electric brakes are suitable for transmitting drivers' intentions to each axle at the speed of electricity. In addition to electric brakes, anti-slip braking systems and electromagnetic rail brake systems should be considered. Another choice in braking systems is wind-pressure braking, which is effective for braking from high speeds of 200 km/h or more. As a result of a wind-tunnel test, as shown in Photo 2.4, the air resistance can be 2 to 2.5 times that of a streamlined vehicle. The lighter the train, the more effective the wind-pressure brake is. At 200 km/h, a braking distance of about

Photo 2.4 Wind-pressure
brake model[3]

2,000 m can be shortened to about 1,600 m, and at 250 km/h, a braking distance of
about 3,000 m can be shortened to about 2,000 m. In the 331-km/h test conducted
in France, engineers first opened all passenger car windows, reduced the speed to
about 200 km/h, and then applied the brakes.

Electric braking absorbs kinetic energy and converts it into heat. It would be
wonderful if this energy could be recovered as a power source in some way.

(7) **Conclusion**

The maximum speed of limited express trains in France is currently 140 km/h, and
SNCF is trying to increase this to 160 km/h. A speed of 200 km/h has not yet been
considered in any country. However, if a new line is constructed based on latest
technologies, speeds of about 200 km/h will have sufficient practical potential,
although many research issues remain.

In 1955, lightweight Series 10 coaches with semi-monocoque structure were
built. Various tests of Odakyu Series 3000 SE cars, which started business operation in July 1957, had already been completed by the time of the commemorative
lecture, so Miki's lecture was based on these achievements.

Among the topics discussed in the previous subsections, Miki had experience
with aerodynamic drag due to high speed, downsizing, reducing vehicle weight,
and lowering the center of gravity. He understood that these matters would prove
essential in designing the Shinkansen vehicles. However, Miki knew that many

---

[3] Miki [4].

unproven matters remained, including the impact of wind pressure when trains pass each other or enter tunnels, the distance between inbound and outbound track, onboard air conditioning, low-air-resistance pantographs, and others.

2.2 Running Safety and Ride Comfort (Tadashi Matsudaira)

Matsudaira spoke on the following six topics.

(1) **Minimum Curve Radius of Tracks**

Centrifugal and gravitational forces act on a car running on a curve. If the direction of the combined force goes outside of the outer rail, the car overturns. To prevent overturning, tracks are canted—that is, they are inclined at curves, as shown in Fig. 2.3.

(2) **Derailment**

When the force acting between the wheel and rail is divided into horizontal force Q and vertical force P, as shown in Fig. 2.4, the larger the Q and the smaller the P, the more easily the wheel gets on the rail. Therefore, the ratio of these two force, Q/P, is called the derailment coefficient. When a derailment occurs, this coefficient is considered to be at least 1.0 or more.

Figure 2.5 shows the derailment coefficient measured on a Series 80 train.

Fig. 2.3 Cant of tracks.
Source Studies of the Tokaido Shinkansen, vol. 1, RTRI, 1960, p. 10

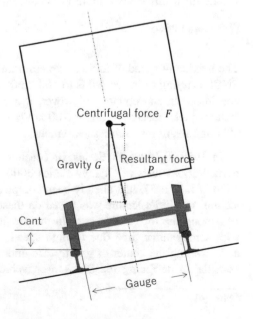

Fig. 2.4 Force between the
wheel and rail. *Source* Studies
of the Tokaido Shinkansen,
vol. 1, RTRI, 1960, p. 11

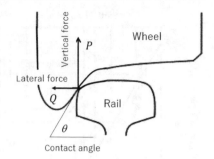

Fig. 2.5 Measured
derailment coefficients.
Source Studies of the Tokaido
Shinkansen, vol. 1, RTRI,
1960, p. 11

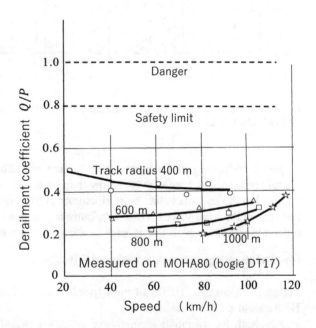

As shown in this figure, the derailment coefficient is sufficiently low in current trains at a speed of 120 km/h, so there is no danger of derailment. However, it is essential to note that the derailment coefficient increases rapidly when the speed exceeds 100 km/h. This is because the Series 80 cars begin to experience hunting motion at this speed. If this hunting motion is prevented in high-speed trains and the trains' vibration is kept sufficiently low from the standpoint of riding comfort, there is no danger of derailment.

(3) Ride Comfort

Figure 2.6 shows the results of many researchers' human body vibration experiments in their investigation of the vibration limits on riding comfort.

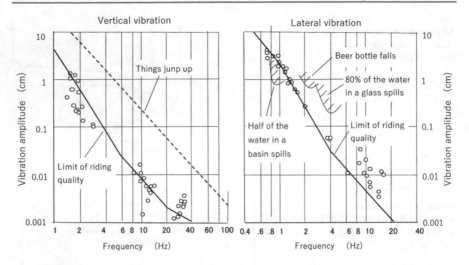

Fig. 2.6 Permissible limits of vehicle vibration. *Source* Studies of the Tokaido Shinkansen, vol. 1, RTRI, 1960, p. 11

Small white dots in the figure show typical vibrations of current coaches and electric railcars. The loud shaking at low frequencies is near the limit of ride comfort values, but the so-called chattering vibrations at high frequencies are several times larger than the limit values. Therefore, current coaches and electric railcars need to reduce vibrations at high frequencies to a fraction of their current level.

(4) Relationship between Speed and Vibration

Figure 2.7 shows the results of high-speed tests conducted between Shizuoka and Hamamatsu on the Tokaido Line.

Although the vibration acceleration increases slightly with speed, it is almost constant above 110 km/h in vehicles designed more recently (e.g., the NAHA10 type). However, based on Fig. 2.7, it is impossible to estimate vehicle vibrations at 200–250 km/h. Doing so requires the use of theory and model testing.

Theoretically, there are two types of vehicle vibration. One is forced vibration due to track irregularities. The other is self-excited vibration, which occurs even if the track is not irregular. The former is expected to become constant at a certain speed, while the latter suddenly appears at a certain speed and becomes more intense, which is the biggest obstacle for high-speed railways. Vertical vibration does not have self-excited oscillation, but left–right vibration has hunting motion, which is a self-excited oscillation. Since the hunting movement is very complicated, it is difficult to estimate it by theory alone, so model experiments are being done to learn more about its nature (Fig. 2.8).

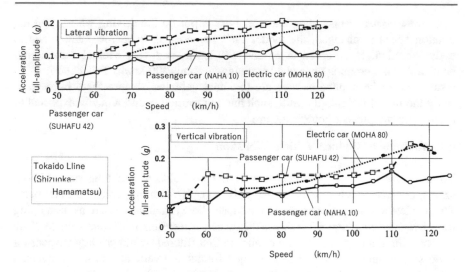

Fig. 2.7 Speed and vibration acceleration. *Source* Studies of the Tokaido Shinkansen, vol. 1, RTRI, 1960, p. 12

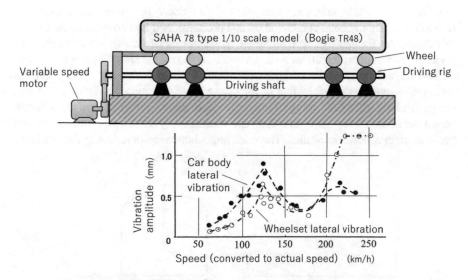

Fig. 2.8 Experiments with roller rig and scale model vehicle. *Source* Studies of the Tokaido Shinkansen, vol. 1, RTRI, 1960, p. 12

In these experiments, when the model car is put on the driving wheels and the rotation speed is changed, the model car's meandering behavior, mainly due to body vibration, is observed at around 120 km/h, and very violent meandering behavior of the bogie is observed beginning at about 200 km/h. This experiment's results cannot be applied as they are to actual vehicles because there is room to study the model's similarity. Still, such hunting motion is undoubtedly expected to occur with the present bogie structure.

(5) Methods to Reduce Vehicle Vibration

To minimize track irregularities, it is highly desirable to use long rails without seams, concrete sleepers with rubber pads, and, furthermore, concrete track beds. The vehicle's spring system should insulate the vibrations caused by remaining track irregularities. Figure 2.9 shows the suspension system of current vehicles. The vertical vibrations coming from the rails are first filtered by the primary suspension (axle spring) and transmitted to the bogie frame; the vertical vibrations are then filtered by the secondary suspension and transmitted to the car body.

Left–right vibrations are filtered by the swing hanger's pendulum action and transmitted to the car body.

The softer the spring, the greater the spring's vibration insulation effect. In short, the reduction of vibration is a matter of how these springs are arranged and how soft they are made while satisfying strength and functional requirements. Theoretical research has already been conducted on this problem, and it is now available for use in design. As for springs, conventional metal springs seem to have reached an impasse, and in order to further reduce vibration, the use of air springs is necessary.

The main feature of air springs over metal springs is that air springs can be softened as much as possible; that is, they have large vibration insulation effect and are particularly effective for high-frequency vibration, and their control devices can adjust the height. They also can correct the vehicle body's inclination on curves and perform other automatic control. The most important thing in reducing the vibration

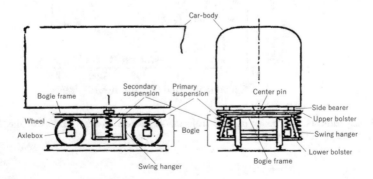

Fig. 2.9 Spring system of current railcars. *Source* Studies of the Tokaido Shinkansen, vol. 1, RTRI, 1960, p. 13

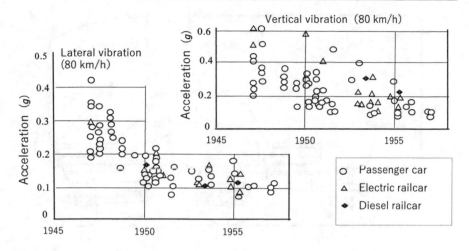

Fig. 2.10 Improvements in vehicle vibration. *Source* Studies of the Tokaido Shinkansen, vol. 1, RTRI, 1960, p. 13

of high-speed vehicles is to avoid hunting. An effective method for doing so is to reduce wheel tread gradients; attach axles to bogie frames firmly enough in the front, back, left, and right direction and apply restoring force or friction by springs to rotational movement of bogies. However, the details of vibration reduction need to be discussed in future studies. The vibration of railcars has decreased by almost one-third from immediately after the war to today (Fig. 2.10). The values for 1947–1948 are of pre-war design bogies, and subsequent values are of new bogies. The decrease in vibration during this period is due to the tracks' improvement and the research on reducing bogie vibration.

(6) Concept of High-Speed Bogie

Figure 2.11 shows a spring system concept for a bogie for 200 to 250 km/h.

It uses partially rubber-filled wheels to reduce impact force on rails, hydraulic air springs as the primary suspension to prevent hunting, and makes the secondary suspension as soft as possible to reduce the car body's natural frequency to one cycle or less. To maintain the lateral stability of the car body, the secondary suspension's position is raised as much as possible by mounting the suspension inside the car body. Car body weight is loaded to the side bearers so that the friction prevents bogie hunting.

Such a bogie will reduce vibration to one-half or one-third of the current vehicle vibration, providing an ideal ride.

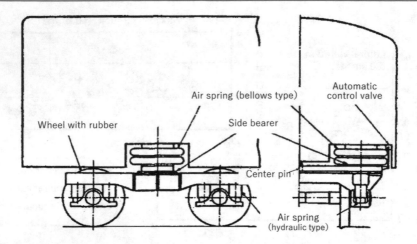

Fig. 2.11 Suspension concept for high-speed railcars. *Source* Studies of the Tokaido Shinkansen, vol. 1, RTRI, 1960, p. 13

Strain gauges made it possible to measure wheel loads and lateral force and thereby to determine the derailment coefficients, Q/P, of running railcars. It was also determined that the cause of the increase in Q/P was hunting, and vehicle vibration had been reduced to about one-third of the conventional level.

Knowing these achievements, the audience must have thought that the current research team would surely be able to realize a safe railway of 200 km/h or more.

2.3 Track Structure (Yoichi Hoshino)

Unless there is something wrong with trains, railway tracks usually have sufficient strength margins that they do not break suddenly at higher train speeds. The problem is the accumulation of tiny amounts of daily damage by trains. However, once tracks are subjected to excessively large lateral force due to hunting motion, they will bend, as shown in Photo 1.17. Naturally, since high-speed rail vehicles must be designed so that hunting does not occur from the viewpoint of safe running and riding comfort, it can be considered that the lateral force acting on the tracks of high-speed rails will not be greater than they are now.

(1) **Train Speed, Progression of Track Irregularities, and Maintenance Volume**

Train vibration is expected to increase at higher speed even if trains run on the same track, and the degree of the vibration is proportional to the train speed or the square of it. Therefore, to keep the train vibration at the same level as the current level, track irregularities should be reduced to 1/ (speed magnification) or 1/ (square of the speed magnification), that is 1/2.5 or 1/6, of the current level when the train speed is increased by 2.5 times. However, the present track irregularities are kept within 5 mm, so they should be reduced to between 1 or 2 mm if the speed is increased by 2.5 times. Meeting this requirement is not an easy task with the current track structure that places sleepers on ballast.

It is known that the function of track beds deteriorates due to the vibrations caused by trains. The degree of deterioration is considered proportional to the vibration acceleration. Since the acceleration increases in proportion to the train speed or the square of it, the track bed's deterioration is eventually proportional to the train speed or the square of it. Therefore, for the same tonnage, if the train speed is 2.5 times faster than the current speed, the track irregularities will progress 2.5 or 6 times faster than the present, and the amount of maintenance required to keep them within 1/2.5 or 1/6 of the current level will be 6 or 36 times greater than the current amount. Therefore, the track bed should be modified in some fundamental way to prevent such a large amount of maintenance.

(2) Track Vibration

Figure 2.12 shows an example of the vibrations that occur on a track when a train is passing.

The upper three waveforms show vertical acceleration of a sleeper, of the track bed (ballast) under the sleeper, and of the roadbed, from top to bottom. The lower three waveforms show lateral acceleration. The rapid vibrations of the sleeper are found to be completely different on the track bed beneath it.

Theoretical studies of these vibration characteristics have been carried out. Figure 2.13 shows how the track bed acceleration changes with the changes in the constants of track's components when a wheelset is dropped onto the track from a certain height.

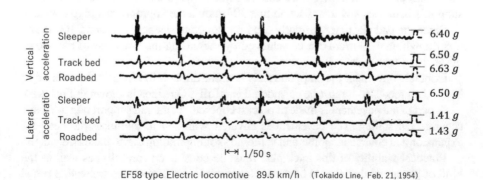

EF58 type Electric locomotive 89.5 km/h   (Tokaido Line, Feb. 21, 1954)

Fig. 2.12 Vibration of each part of track. *Source* Studies of the Tokaido Shinkansen, vol. 1, RTRI, 1960, p. 14

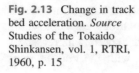

Fig. 2.13 Change in track bed acceleration. *Source* Studies of the Tokaido Shinkansen, vol. 1, RTRI, 1960, p. 15

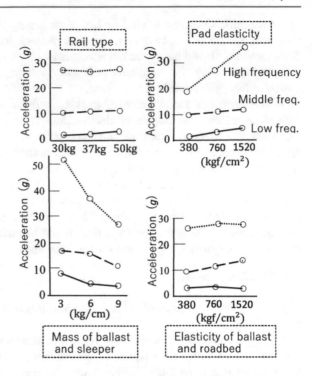

Figure 2.13 also shows that the vibration isolation of the track bed is achieved by either softening rail pads or making the part below the rails heavier.

If the rubber pads softly support the rails, it can be expected that the part below the rails can perhaps be a hard structure like concrete pavement.

(3) Proposal of New Track Structure

For a track paved with concrete, a theoretical study was conducted to determine the rail support's hardness, and it was found that 30 tf/cm is the appropriate spring constant.

Since the pad elasticity currently used for concrete sleepers is around 100 tf/cm, it is thought that 30 tf/cm can be achieved by increasing the thickness of the pad or by layering two pads.

Concrete pavement that supports rails is more difficult to repair in the event of a break than roads, so it must be considered carefully. One idea is shown in Fig. 2.14.

A part of the pavement is made of precast concrete for the rail support part to enable accurate rail setting. The pavement is covered with soil to eliminate the need for expansion and contraction joints and to provide soundproofing due to increased weight.

Practical viability of this track plan must be considered carefully, as well as the kind of pavement that should be used. Nonetheless, for high-speed trains that run at 2 to 2.5 times the current speed, it is expected that the current track structure is entirely ineligible.

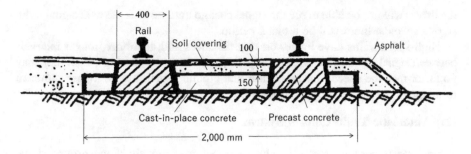

Fig. 2.14 New track structure. *Source* Studies of the Tokaido Shinkansen, vol. 1, RTRI, 1960, p. 16

Recent track dynamics has enabled design that incorporates the most problematic track dynamic loads and roadbed soil that support pavements. It seems that the tracks shown in Fig. 2.18 have already been proposed and promoted in foreign countries, so I want to know how durable this type of track is against trains' dynamic loads. Using the recently completed testing machine, I want to obtain a practical perspective as soon as possible.

Hoshino's lecture was based on the track dynamics described in Sect. 1.3, with the use of long rails. For railways traveling at over 200 km/h, Hoshino expected difficulty maintaining the conventional ballast track and proposed a new concrete track structure.

2.4   Signaling Safety (Hajime Kawanabe)

The distance from brake application to stop is called braking distance. If the braking distance of a train running at 45 km/h is 80 m, that of a train running at 250 km/h will be 2.5 km, about 30 times that at 45 km/h. This distance is based on the assumption that deceleration is a certain value, and it is not yet known how much deceleration can be achieved with a high-speed train of 250 km/h.

(1)   Disadvantage of Ground Signals

Tri-color signal lights can be seen from a distance of 1,000 m when conditions are good, but if it rains or there is fog, the driver may not be able to see them even from a distance of 200 m. Even in that case, trains running at 45 km/h can stop before the signal light if the driver applies the brakes after seeing the red light. However, since a train running at 250 km/h has a long braking distance, signal lights must be seen from 3 to 4 km away in order to stop in time. Moreover, if the driver of a 250-km/h train recognizes a signal 800 m away, the driver has about 11 s to see it, but if the visibility distance is 200 m due to bad weather, the signal is only visible for about 3 s. Also, if the signal changes immediately after the train passed the signal light, the driver will not be able to see the signal change until he sees the next signal light. In other words, there will be a blank period.

High-speed trains have longer braking distances and, therefore, longer intervals between signal lights, and thus longer periods of blankness, during which they may go faster than the speed limit. It is clear that ground signals are no longer sufficient to maintain the safety of high-speed trains.

(2) **Automatic Train Control System**

A new system contains a cab signal system that continuously shows ground signal lights to the driver. With this system, changes in ground signals are transmitted to the car no matter where the train is running, and the driver can immediately adjust the speed. However, even if such a cab signal is installed, the driver could miss the signal due to carelessness. Therefore, when the signal changes to a dangerous side, such as from green to yellow, an alarm will be issued to arouse driver's attention.

A device that automatically applies the brakes if the driver does not use the brakes is called an automatic train control system and is essential for high-speed trains such as those running at 250 km/h. Figure 2.15 shows the principles of this system.

The onboard equipment receives the signal current sent out from the transmitter on the ground to the rails and displays the same signal as the ground on the car. It also measures the train's speed and automatically applies the brakes if the speed exceeds the limited speed indicated by the onboard signal.

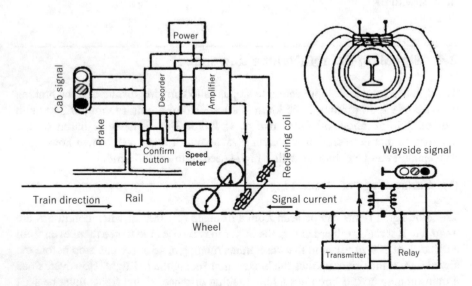

Fig. 2.15 Principles of automatic train control system. *Source* Studies of the Tokaido Shinkansen, vol. 1, RTRI, 1960, p. 18

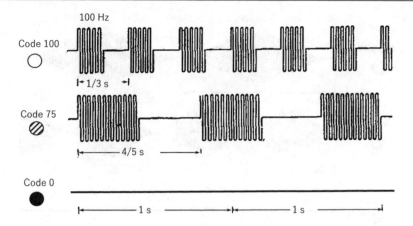

Fig. 2.16 Signal current patterns in America. *Source* Studies of the Tokaido Shinkansen, vol. 1, RTRI, 1960, p. 19

The signal current waveforms flowing through the rails are as shown in Fig. 2.16 (signals used in the USA). In the case of green (proceed signal) and yellow (caution signal), a current of 100 Hz flows intermittently, as shown in the upper part and center of the figure; and for the red signal, the current does not flow, as shown at the bottom of the figure.

JNR has been studying a continuous cab warning system to be installed on the Tokaido Line in March of 1958. In this system, as shown in Fig. 2.17, the signal current of 1,300 Hz is increased or decreased at a rate of 20 times per second for the green signal (proceed signal) and at a rate of 35 times for the yellow (caution signal). No current is applied for the red (stop signal), as shown at the bottom of the figure.

The system rings a buzzer or bell when the signal changes to a dangerous side but it has not reached the stage of automatic braking. A vacuum tube oscillator is used for the transmitter on the ground, the output is 4 W or more, and the receiving coil on the car senses the signal when a current of 0.05 A flows through the rails. This system is designed for ground signals with intervals of up to 2 km. For high-speed trains, the system should be designed for longer signal intervals. This issue needs further study.

(3) Signal Systems for High-Speed Rails

The method shown in Fig. 2.17 was originally developed for a cab signal system for AC electrified lines, so it can be used for both AC and DC electrified lines. However, since the signal frequency and onboard equipment are designed for DC electrified lines, they must be modified for use with AC electrified lines; this is a significant research topic for the future. In AC electrified lines, large interference voltage with distorted waveforms appears in the on-board receiving coil and interferes with the signal voltage.

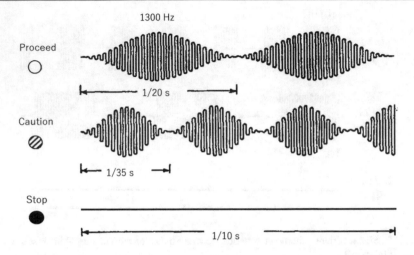

Fig. 2.17 Signal current. *Source* Studies of the Tokaido Shinkansen, vol. 1, RTRI, 1960, p. 19

(4) **Closing**

Currently, it is not possible to design an actual signaling system because traffic volume, number of stops, type and number of trains, and operating systems are not fixed.

Ideally, in the future, the following train should automatically measure the speed of the preceding train and the distance to the preceding train, compare the speeds of the two, and automatically maintain a constant distance. Alternatively, many trains could be automatically operated by radio from the control center, but this way is still technically challenging and not feasible at the moment. This is a next research task.

Kawanabe's lecture was based on the AF track circuit and the cab warning system described in Sect. 1.4. The last part of the lecture is especially noteworthy: "Many trains could be automatically operated by radio from the control center, but this way is still technically challenging and not feasible at the moment." Nonetheless, the world's train control systems are currently moving in this direction, and the first such system in Japan was implemented in 2011 under the name ATACS (short for Advanced Train Administration and Communications System).

The fiftieth anniversary commemorative lecture was a success. In his book, Shinohara wrote [2]:

I saw that the audience was extremely attentive. From their response, I was convinced that a new high-speed railway could certainly be realized. The response to the lecture was great. Prominent newspapers, including the Asahi Shimbun, which sponsored us, gave us very favorable coverage of our project. And this aroused public opinion. (Author's translation of the Japanese)

Photo 2.5 Then JNR president Sogo (Author)

Shinohara commented on JNR President Sogo's reaction as follows [2]:

President paid attention to the content of the lecture and told me to hold another special briefing for himself, Vice President, and other JNR executives. He was quite satisfied with our explanations and said that he would be the one to lead the effort to realize the plan. (Author's translation of the Japanese)

Thus, President Sogo[4], with public opinion on his side, gave concrete shape to the Shinkansen project (Photo 2.5).

References

1. Tadashi Matsudaira, "A Retrospective of Research and Development on the Tokaido Shinkansen (in Japanese)", *Journal of the Japan Society of Mechanical Engineers,* no. 646, 1972, p. 101.
2. Takeshi Shinohara, Hideshige Takaguchi, *The Soliloquy of the Shinkansen Inventor* (in Japanese), Pan Research Institute, Japan, 1992, p. 82.
3. Takeshi Shinohara, "Greetings on the RTRI commemorative lecture (in Japanese)", Research on the Tokaido Shinkansen, vol. 1, RTRI, 1960, p. 6.
4. Miki, T., "Problems Related to High-Speed Railway," *Journal of the Japan Society of Mechanical Engineers,* no. 480, 1959.

---

[4] Shinji Sogo joined the Railway Bureau in1909. He became the fourth JNR president in 1955.

The Shinkansen Line: Decision to Construct the Line and the Remaining Technological Problems

3

3.1 Outline of the Decision Process for Building the Shinkansen

Ten years after the end of the war, Japan was recovering from its defeat, and the economy was beginning to grow. As a result, there was a shortage of transport capacity on the Tokaido Line, the most crucial trunk line in Japan. So, in May 1956, JNR established a study committee to deliberate on measures for solving the problem. The committee discussed several plans, including:

- A plan to add a new narrow gauge line parallel to the existing Tokaido Line.
- A plan to build a new narrow gauge line regardless of the existing line.
- A plan to build a new standard gauge line regardless of the existing line.

The committee met five times. President Sogo and Chief Engineer Shima supported the standard gauge plan, but most of the committee members supported the narrow gauge plan; unable to reach a consensus, the committee adjourned. However, the RTRI's anniversary lecture, which was held four months after the committee deadlock, became the catalyst for remedying this situation.

Two months after the memorial lecture, in July 1957, JNR requested the government to settle the matter from a national perspective, which would include air and sea transport. The government set up a committee in August 1957 and, in March 1958, concluded that it was appropriate to build a new standard gauge line. In 1975, Shima recalled in the *Nihon Keizai Shimbun* newspaper [1]:

> There were many twists and turns before the decision was made to build a new standard gauge line. Despite our persuasion, many said it was risky to adopt a new gauge that was incompatible with the current rail network. However, President Sogo was very determined to build high-speed rails with the standard gauge. (Author's translation of the Japanese)

© The Author(s), under exclusive license to Springer Nature Singapore Pte Ltd. 2022
T. Shimomae, *Birth of the Shinkansen*,
https://doi.org/10.1007/978-981-16-6538-7_3

3.2 Technical Issues to Be Resolved

High-speed railways were "a possibility" until the commemorative lecture, but once the decision was made to go ahead with construction, RTRI was tasked with creating a railway that would be safe, stable, and able to withstand practical use at over 200 km/h within a limited amount of time. Therefore, RTRI organized eight research groups with 173 research themes, as shown in Table 3.1, and started full-scale technological development work toward making a modern high-speed railway a reality. The year 1961 was to be devoted to the design and production of prototype trains, and running tests on a test section were scheduled to start in 1962, so it was imperative that main issues be resolved in the three years between 1958 and 1960.

In his book *Tokaido Shinkansen*, Ryohei Kakumoto, who was a member of the Shinkansen planning division, wrote about the period from the construction start to the opening of the Shinkansen as follows [2]:

> It was unprecedented in railway construction history that the most modern double-track railway was completed over 500 kilometers in five years while developing new technology. It was a major undertaking whose success was still in jeopardy, even with the JNR's concerted efforts. Five years was too short a period of time since the technical specifications of the railway had not been determined. However, considering that the Tokaido Line would face serious transportation difficulties in a few years, the project had to be completed, no matter how difficult it was. An extended period of research would be necessary to produce something excellent. However, the construction work had to start soon, or else it would not

Table 3.1 Research groups and research subjects

Group	Main subjects	Number of research themes
(1) Track structure	Tracks, Roadbeds, Rail fastening devices, Rail, Sleepers	25
(2) Design of high-speed railcars	Aerodynamics, Power transmission, Main motors, Car bodies, Bogies, Axles, Bearings	29
(3) Dynamics of high-speed railcars	Hunting prevention, Suspension, Riding quality	17
(4) Braking	Braking methods, Adhesion characteristics	18
(5) Power collection	Overhead catenary equipment, Catenary clamps, Dynamics of power collection system, Pantographs	25
(6) AC electrification	Track circuits, Cab signals, Power feeeding, Electric induction	18
(7) Signal systems	Introduction of electronic devices to signaling systems, ATC, CTC, Moving block system by radio waves	19
(8) Automatic train operation	Methods of automatic train operation, Automatic train rescheduling, Programmed control of CTC	22
Total number		173

Source Research on High-speed Railways, Kenyusha Foundation, 1967, p. 620

complete in time. We arranged to allocate the limited five-year period to research and construction, so that the research work would not interfere with the construction work and the construction work would be done safely but as quickly as possible to allow the necessary time for research. (Author's translation of the Japanese)

Until test runs on the test section began, various tests were conducted using conventional lines. The number of the tests on conventional lines was 56, the number of the tests on the test section was 86 with two prototype trains (Train A and Train B), and 33 with one six-car pre-mass-production type (Train C), a total of 175 tests. The record of these studies by the RTRI is enormous, and it is not easy to summarize them. In this book, the author divided these extensive studies into three categories including nine research topics.

Category 1

This category includes phenomena that did not exist in conventional railways. Hunting of bogies and large contact loss of power collection systems fall into this category. These problems cannot be solved by only making conventional ones stronger or more precise. The construction and verification of a theory that can explain the phenomenon are essential; otherwise, one would have to go through a trial-and-error process, which is unlikely to yield results.

As mentioned in Sect. 1.1, the hunting motion problem had been studied even before the Shinkansen project started. This was a decisive factor in the Shinkansen's opening in 1964.

On the other hand, as seen in Sect. 1.5, the research history of high-speed power collection systems was short. As will be described in Sects. 4.2 and 4.9, it can be said that the Shinkansen's power collection system was put into practical use with problems that could not be solved on the test section.

Category 2

This category includes problems that have traditionally existed but increase with higher speeds. These include track structure, aerodynamic drag, vehicle weight reduction, vehicle vibration (ride quality), power energy supply, and many other topics. This category's challenge is to understand or estimate how much the problem will intensify with the increase in speed and to take countermeasures such as changing the shape or size of the system as well as strengthening the system.

Category 3

This category includes matters that must be addressed because it is not possible to operate or manage manually with the increase in speed. New systems must be designed and implemented to solve the problem. The automatic train control system (ATC) and the centralized traffic control system (CTC) fall into this category.

References

1. Hideo Shima, "My resume (in Japanese)," *Nihon Keizai Shimbun,* April 28, 1975.
2. Rhouhei Kakumoto, *The Tokaido Shinkansen* (in Japanese), Chuko Shinsho, Japan, 1964, p. 14.

Technical Development Necessary for Realizing the Shinkansen

4

4.1 Overcoming Hunting Motion

Before getting to the main topic of this chapter, let us review Matsudaira's research on hunting motion:

- In 1947, two years after he came to RTRI, Matsudaira began his research on railcars' hunting behavior. In December of the same year, he presented an analysis of hunting motion that took into account all the force acting on a wheelset, including creep force.
- In 1949, he completed a roller rig for a scale model wheelset and verified his analysis via experiments using this model. This was the first use worldwide of roller rig testing for vehicle dynamics.
- In his analysis model at the time, axles were rigidly attached to bogie frames. Still, he noted that the coupling between axles and bogie frames was not fully rigid in reality, so he analyzed the case where axles were elastically attached. This was a crucial step in overcoming hunting motion and was the first analysis of its kind.
- His research improved the ride quality of new trains produced after 1950 for conventional lines and enabled freight trains to travel at higher speeds. However, as analytical models became more complex, it became difficult to express the hunting critical speed v_c in the form of $v_c = f(\ldots)$, so for 2-axle freight cars, he had to use graphical methods to obtain v_c. With bogies having primary and secondary suspension and swing hanger systems, analysis became too complicated to calculate manually. Therefore, his study method shifted mainly to model experiments using roller rigs.
- In 1956, 1/10 scale model's rolling tests showed that primary hunting occurred at real speeds of 100–150 km/h, and secondary hunting occurred at speeds above 200 km/h. With the bogie structure of the time, secondary hunting proved to be almost inevitable. Therefore, Matsudaira decided to investigate the phenomenon in more detail by enlarging model's scale to 1/5th (Photo 4.2).

© The Author(s), under exclusive license to Springer Nature Singapore Pte Ltd. 2022
T. Shimomae, *Birth of the Shinkansen*,
https://doi.org/10.1007/978-981-16-6538-7_4

Photo 4.1 a Rolling
experiment with 1/10 model.
(Provided by RTRI).
b Enlarged of (Photo 4.1a)
(Provided by RTRI)

The preceding summary depicts the progress of hunting motion research up to
the RTRI's commemorative lecture. Let us now trace the history of hunting motion
research from the memorial lecture to the Shinkansen opening.

4.1.1 Experiments with 1/5 Model Bogie (One Bogie, 1959)

Matsudaira completed a 1/5 model rolling rig for a single bogie (Photo 4.2) and
conducted experiments to show the relationship between hunting motion and various factors, including the wheelbases, lateral elasticity between the axles and the
bogie frame, and loading ratios of the body weight to the side bearers and the center
pivot. The results were as follows:

(i)   The critical speed at which hunting begins is largely independent of the body
      weight (Fig. 4.1a).
(ii) The critical speed increases as the wheelbase increases. When the center
      pivot supports the body weight, the increase rate is roughly consistent with
      the results determined by Eq. 1.2 in Sect. 1.1 (Fig. 4.1b).
(iii) When the side bearers support the car body weight, the side bearers' friction
      will significantly increase the critical speed (Fig. 4.1c).

Photo 4.2 Rolling test of 1/5 scale model. (Provided by RTRI)

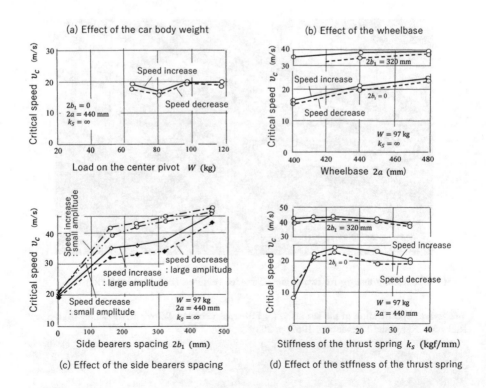

Fig. 4.1 Relationship between each factor and hunting motion. *Source* Studies of the Tokaido Shinkansen, vol. 2, RTRI, 1961, p. 218

(iv)   When the center pivot supports the body weight, the critical speed becomes
        maximum at a certain value of the elasticity between the axles and bogie
        frame. However, when the side bearers are supporting the body weight, the
        effect is not clear (Fig. 4.1d).

In the experiments with two bogies shown in Photo 4.3, the critical speed
variation's tendency was almost the same as the results shown in Fig. 4.1.

In his report, Matsudaira wrote [1]:

From these experiments, it has been concluded that the secondary hunting can be prevented
up to a maximum speed of 200 km/h by providing appropriate amount of side bearers'
friction and support elasticity to axles, without special measures such as the use of independent wheels. (Author's translation of the Japanese).

At this stage, a framework for preventing hunting motion in high-speed railcars
was established, although specific values had not yet been determined.

At the end of August 1959, the long-awaited test stand for real railcars, which
had been planned for some time, was completed. Table 4.1 shows the specifications
of the test stand.

Photo 4.3 Rolling test rig for two-bogie 1/5 model vehicles. (Provided by RTRI)

Table 4.1 Specification of the test stand for full-scale vehicles. *Source* Research on High-Speed
Railways, Kenyusha Foundation, 1967, p. 288

Guage	1,435 mm (variable from 1,000 to 1,676 mm)
Wheelbase	2,500 mm (variable from 1,500 to 5,500 mm)
Roller diameter	1,060 mm
Rotational speed of the roller	0–1,250 rpm
Vehicle speed	0–250 km/h

4.1.2 Rolling Test of Full-Size Experimental Bogie (One Bogie, 1960)

Photo 4.4 shows the full-scale experimental bogie designed to change various factors.

Photo 4.5 shows the test stand and the experimental bogie loaded with the mass equivalent to half of the car body during rolling tests.

The aforementioned various factors of the bogie include tread gradients (three types), axle box suspension (three types), lateral axle elasticity (three or four stages), body weight sharing ratios of the center plate and side bearers (two stages), axle spring damper constants (two steps). Three types of axle box suspension were: Minden-Deutz (leaf spring type, from now on abbreviated as Minden), Alsthom type, and Schlieren type. The axle box suspension is a device that holds an axle box to allow it to move freely in the vertical direction and slightly in the lateral direction but not in the front-rear direction.

Minden is very stiff in the front-back direction, whereas Schlieren and Alsthom are structurally slightly elastic. In June 1961, the experimental bogie was placed on the test stand, and the world's first full-scale, actual-speed rolling test began. Over the next year, Matsudaira repeated hunting experiments up to 250 km/h using this test stand and the experimental bogie, verified the aforementioned research results that used scale models and confirmed each factor's influence on hunting. Based on the results, he determined the specifications of the prototype bogies for the test section.

Figures 4.2, 4.3, and 4.4 show the results when using the Minden axle box suspension, which was prone to hunting motion.

Photo 4.4 Full-scale experimental bogie. (Provided by RTRI)

Photo 4.5 Experimental
bogie during rolling tests.
(Provided by RTRI)

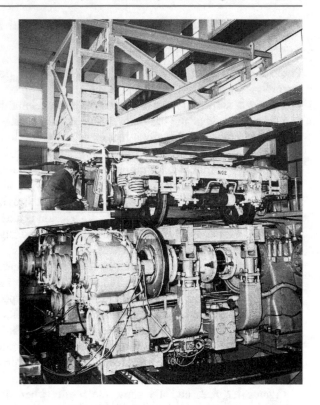

Figure 4.2 shows the results when the body weight is fully placed on the side
bearers, the tread slope is 1/40, and the axle lateral elasticity is 12 kgf/mm. Under
these conditions, primary hunting of about 1.7 Hz occurred at a speed of about 130
km/h, grew to about 160 km/h, and remained nearly constant beyond that. No
secondary hunting occurred.

The curve on the left in Fig. 4.3 shows how the axle's lateral elasticity of the
primary suspension affects the critical hunting speed (tread gradient: 1/40). It shows
that the hunting critical speed is maximized when the axle lateral elasticity is about
100–150 kgf/mm. The right side of Fig. 4.3 shows that the axle spring damping
also has an optimum value that narrows the unstable zones.

Figure 4.4 shows the results when the body weight is fully loaded on the center
plate and wheels with a tread slope of 1/20 are mounted. In this case, the primary
hunting occurred at about 120 km/h and disappeared at about 200 km/h, while the
secondary hunting appeared at about 240 km/h and increased rapidly, changing the
frequency from 2.7 Hz to 4.5 Hz.

In general, the situation worsened as the wheels' tread shape changed due to
wear, and hunting became more severe, especially in the case of a bogie equipped
with Minden. However, in the case of an Alsthom-equipped bogie, when side
bearers supported the body and the lateral axle elasticity was increased to 350–500
kgf/mm, hunting did not occur until speeds of 250 km/h.

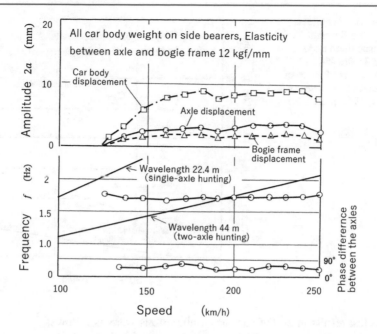

Fig. 4.2 Relationship between speed and hunting (Axle box suspension: Minden; bodyweight support: side bearers, tread gradient: 1/40; axle lateral elasticity: 12 kgf/mm). *Source* Studies of the Tokaido Shinkansen, vol. 2, RTRI, 1961, p. 222

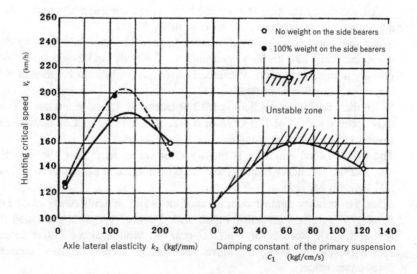

Fig. 4.3 Relationship between hunting critical speed, axle lateral elasticity, and axle spring damping constant (tread gradient: 1/40). *Source* Studies of the Tokaido Shinkansen, vol. 2, RTRI, 1961, p. 224

Fig. 4.4 Hunting critical speed with full weight on the center plate (tread slope: 1/20, lateral axle elasticity: 12 kgf/mm). *Source* Studies of the Tokaido Shinkansen, vol. 2, RTRI, 1961, p. 222

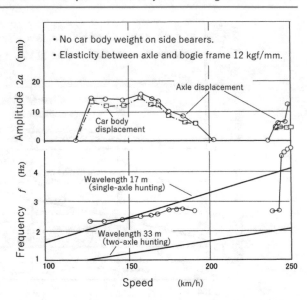

The test results in all the various combinations were as follows:

(i)   Primary hunting can be prevented by lowering the natural frequency of the body rolling, giving damping to this movement, and giving elasticity to the axle support.

(ii) When the side bearers support all or part of the car body weight, the speed at which secondary hunting starts can be increased considerably, and hunting amplitude can be suppressed even if hunting occurs.

(iii) When the wheel tread gradient is reduced to 1/40, secondary hunting does not occur until 250 km/h. Depending on the body and bogie conditions, primary hunting may occur at relatively low speeds, but can be prevented by the actions mentioned in item (i).

(iv) When the wheel tread gradient is 1/20 or treads are worn, secondary hunting may occur at speeds of 170 km/h or more. This hunting motion is extremely severe, so it must be prevented.

(v)   The axle's lateral elasticity is mainly effective in increasing the initiation speed of the secondary hunting, and its proper value is estimated to be 300–500 kgf/mm per axle.

(vi) Since the stability against hunting motion is higher with Alsthom or Schlieren type axle box suspension than with Minden type, it is presumed that giving the axle elasticity in the front-rear direction is also effective in suppressing hunting. However, more research is needed to determine the appropriate value.

(vii) Based on the foregoing statements, even a bogie of the structure in current use is expected to be able to prevent the hunting motion in the range of 250 km/h or less, as long as wheel tread slopes, friction against bogie rotation, and axle support elasticity are adequately selected.

Through these experiments, Matsudaira quantitatively confirmed the hunting motion knowledge that he had gained through analysis and model experiments. At the same time, however, a new problem, which he had not considered, emerged: the effect of axle elasticity in the longitudinal direction. The effect of lateral axle elasticity on hunting motion had been known for some time, but the effect of longitudinal axle elasticity had not been studied in detail. These experiments of the full-size test bogie just happened to show Matsudaira its importance.

4.1.3 Hunting Motion Analysis with Extended Conditions (One Bogie, 1961)

Based on the aforementioned results, Matsudaira set out to analyze the effects of suppressing hunting due to the elastic support of axles in the front-rear direction and restoring springs and dampers that act against bogie rotation.

Using the analytical model shown in Fig. 4.5, Matsudaira formulated the equations for the equilibrium of the force and rotational moment for wheelset AB, wheelset CD, and the bogie frame. Then he solved the equations using graphical methods, which took a lot of time and effort. At the time, RTRI had its first computer, a Bendix G-15, but it could not perform his numerical analysis.

4.1.3.1 Front-Rear Elasticity Between Axle and Frame (k_1 in Fig. 4.5)

Matsudaira wrote in his 1961 research report as follows [2]:

> It has been known from our calculations of the hunting motion of two-axle wagons that the elasticity value of axles to the bogie frame in the lateral direction has an appropriate value that maximizes hunting critical speed. However, this was only for a special case where the support elasticity of axles is infinite in the front-rear direction. Therefore, in order to establish design guidelines for the axle box suspension of the Shinkansen bogie, I will take up again this issue and study the appropriate values of the elasticity in the front-rear and lateral directions. Specifically, using the dimensions and weight of the Shinkansen vehicle, I will calculate the hunting characteristics when the axles are firmly attached to the bogie frame longitudinally to change the left–right elasticity, and when the axles are firmly attached to the left–right direction to change the front-back elasticity. (Author's translation of the Japanese).

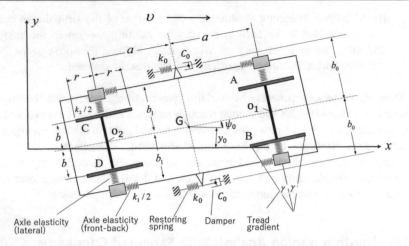

Axle elasticity   Axle elasticity   Restoring   Damper   Tread
(lateral)   (front-back)   spring   gradient

Fig. 4.5 Analytical model of restoring springs and dampers acting against bogie rotation. *Source* Studies of the Tokaido Shinkansen, vol. 3, RTRI, 1962, p. 257

The results were as follows:

(i)   When the front-rear elasticity of the axles is made infinite, there is a lateral elasticity value that maximizes the hunting critical speed.
(ii)   When the axles' lateral elasticity is infinitely large, there is also a front-rear elasticity value that maximizes the hunting critical speed.
(iii) Comparing the two, the latter is about 50% larger than the former.

That is, the result was that fixing the axles to the left-right direction and giving elasticity to the front-rear direction had a higher hunting critical speed than vice versa. About this, Matsudaira said [3]:

As a practical matter, it is impossible to make the left and right elasticity infinite, but it will not be so difficult to make it about $k_2 = 2000\text{kgf/mm}$. The appropriate value of the front-back elasticity for this is $k_1 = 680\text{–}1020$ kgf/mm. In other words, in the case of the Shinkansen bogie, the design value of the axle box support elasticity should be about 1000 kgf/mm in the left–right direction and 350–500 kgf/mm in the front-rear direction per axle box. In this case, the hunting critical speed will be 280 km/h. (Author's translation of the Japanese).

4.1.3.2 Appropriate Values for Springs and Dampers to Suppress Bogie Rotation (k_0 and c_0 in Fig. 4.5)

It has been confirmed in the 1/5 scale model tests that the hunting critical speed can be increased by loading a part of the body weight on the side bearers. However, from scale model experiments, it was impossible to obtain values that should be reflected in designing real cars. Therefore, using the analytical model shown in

Fig. 4.5, Matsudaira quantitatively evaluated the effect of k_0 and c_0 in the figure. The results were as follows:

(i) When a restoration spring with rigidity of 300 kgf/mm is used, the hunting critical speed becomes 1.5 times higher than that without the spring, and when the rigidity is 670 kgf/mm or more, hunting does not occur. In this way, the restoration springs are very useful for suppressing hunting. However, since the lateral force acting on rails becomes large, it is recommended that bogie's rotation be suppressed by using the friction of side bearers or large center pivots.

(ii) When a 35 kgf/cm/s damper is used, the hunting critical speed can be increased up to 1.5 times without the damper. This damping effect can be realized with oil dampers, but from the bogie structure's viewpoint, it is still desirable to use friction with side bearers or large center pivots.

Based on these determinations, in November 1961 a two-car train (Train A) and a four-car train (Train B) were ordered. The axle box suspension was designed to be elastic in both the longitudinal and lateral directions.

4.1.4 Rolling Test of Prototype Vehicle (Car Body with Two Bogies) and Analysis of Its Motion (1961–1962)

Based on the experimental bogie's actual speed rolling tests, six types of bogies for test vehicles were completed. In combination with car bodies, rolling tests up to 250 km/h began in February 1962. Since each bogie's wheel tread gradient was as small as 1/40, secondary hunting did not occur, but primary hunting appeared. When suspension's damping values were selected appropriately, hunting could be eliminated in some cases but could not be eliminated in some cases.

Therefore, the following hunting motion analysis of a vehicle (vehicle body with two bogies) was carried out. Matsudaira described the purpose of the analysis as follows [4]:

Many theoretical calculations and model experiments have been carried out, as well as hunting motion tests on the test stand of the actual experimental bogie, to find hunting prevention methods for designing the Shinkansen bogie. However, the conventional hunting motion theories consider only a bogie and not a whole vehicle. Moreover, the solutions only find the critical speed of the hunting motion and do not give any knowledge about the stability of vehicle vibrations at other speeds. Therefore, it became necessary to theoretically elucidate the vehicle's hunting motion of a whole railcar to interpret and generalize the results of the hunting phenomenon of an actual vehicle at the real-speed test stand and draw useful conclusions for bogie design.

Thus, I calculated a railcar's hunting motion by using a new and full-scale method. The method is to formulate equations of lateral motion of a vehicle running on a track, derive the characteristic equation, find its roots by using an electronic computer, and investigate the motion's stability with the roots. This allows us to find the values of the frequency and the damping rate (or divergence rate) of all types of vibrations that can occur in a vehicle at given speeds, and thus to know the state of the vehicle's lateral motion. (Author's translation of the Japanese).

Previously, vehicle body's influence on hunting motion was analyzed by loading half the mass of a body onto a single bogie, because one bogie was the limitation of manual calculations. However, since it became possible to test one railcar on the test stand, and test runs began on the test section, it became necessary not only to analyze the speed at which hunting motion occurs but also to analyze overall motion of "a car body with two bogies" for a wide range of speeds. In 1962, RTRI's Bendix G-15 computer was replaced by a new G-20 transistorized computer, and Matsudaira's new analysis became possible.

Calculations were performed on the models shown in Figs. 4.6, 4.7, and 4.8.

Since the axles were assumed to be rigidly attached in both the lateral and front-rear directions, there were seven types of motion: lateral movement and yawing of the front bogie, lateral movement and yawing of the rear bogie, lateral movement, rolling, and yawing of the body.

The unknown quantities are the lateral displacements of the front and rear bogies and their angular displacements around the z-axis (i.e., y_{T1}, ψ_{T1}, y_{T2}, ψ_{T2} in Fig. 4.6) and the lateral displacements of the body and its angular displacements around the z-axis and x-axis (i.e., y_B, φ_B, ψ_B in Fig. 4.6). In the suspension models

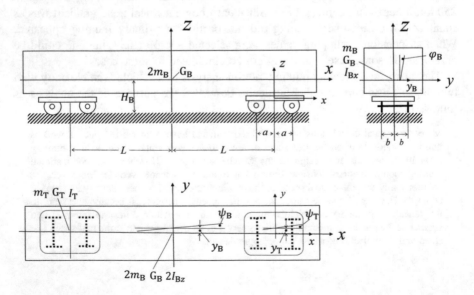

Fig. 4.6 Vibration analysis model for the Shinkansen vehicle. *Source* Studies of the Tokaido Shinkansen, vol. 4, RTRI, 1963, p. 221

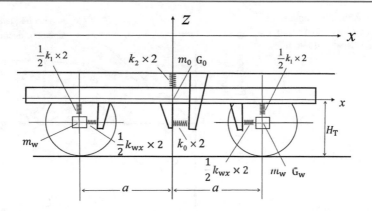

Fig. 4.7 Suspension model for the Shinkansen bogie. *Source* Studies of the Tokaido Shinkansen, vol. 4, RTRI, 1963, p. 221

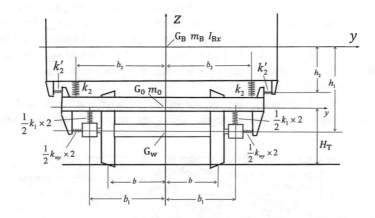

Fig. 4.8 Suspension model for the Shinkansen bogie. *Source* Studies of the Tokaido Shinkansen, vol. 4, RTRI, 1963, p. 221

shown in Figs. 4.7 and 4.8, the swing hanger system, which absorbs lateral vibrations, is replaced by the lateral elasticity of air springs used in the secondary suspension and is represented by k_2' in Fig. 4.8.

Matsudaira obtained the roots to the characteristic equation derived from motion equations using the G-20 computer. In this case, the seven roots represent vehicle's seven vibration modes, but two of them were negative numbers which are not related to hunting.

As shown in Fig. 4.9, one solution represents one vibration mode, the horizontal axis represents the speed, and the vertical axis shows the real part of the solution and the angular frequency. In the figure, the real part of the solution, α, is negative from 0 to v_{c1}, positive between v_{c1} and v_{c1}', and negative again over v_{c1}', so that

Fig. 4.9 Vibration modes
represented by a solution to
the characteristic equation

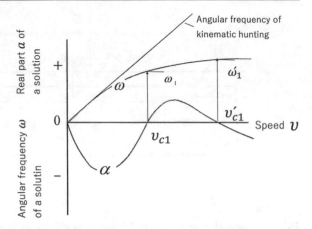

hunting occurs between v_{c1} and v'_{c1} , and the corresponding angular frequencies are ω_1 and ω'_1, respectively.

Figure 4.10 shows the relationship between the spring's strength k_0, which gives the bogie rotation restoring force, and the hunting critical speed v_c, obtained from the five hunting modes of a Shinkansen vehicle equipped with wheels with a tread gradient of 1/40. The tongue-like curves protruding from left to right in the figure are rolling with a low rotation center, rolling with a high rotation center, and yawing by body hunting, from bottom to top.

When k_0 is 100 kgf/mm, increasing the speed indicates that upper-center rolling begins at about 45 m/s, yawing begins above 60 m/s, upper-center rolling ends

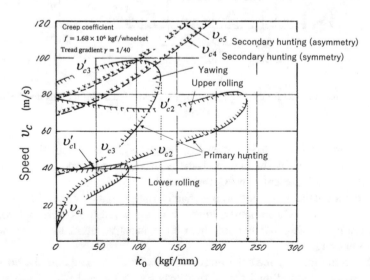

Fig. 4.10 Relationship between k_0 and hunting critical speed. *Source* Studies of the Tokaido Shinkansen, vol. 4, RTRI, 1963, p. 230

above 70 m/s, bogie hunting motion begins at about 90 m/s, yawing ends before 100 m/s, and another mode of bogie hunting begins. It can also be seen from Fig. 4.10 that lower-center rolling does not occur at k_0 above about 90 kgf/mm, yawing does not occur at k_0 above 130 kgf/mm, upper-center rolling does not occur at k_0 above 240 kgf/mm, and all hunting motion ceases when k_0 is made stronger.

Matsudaira outlined the conclusions from this calculation as follows [5]:

(i)   When the restoring moment against bogie rotation is small, or the spring system's damping is insufficient, primary and secondary hunting motion generally occurs.
Primary hunting is classified into three types: lower-center rolling, upper-center rolling, and yawing. Secondary hunting is classified into two types, with the front and rear bogie in-phase and in opposite phases.

(ii)   The instability of the primary hunting is relatively small and may become stable in some cases. However, the secondary hunting's instability increases sharply beyond the critical speed and does not decrease again.

(iii)   To prevent hunting motion, it is necessary to give restoring force above a certain value to bogie rotation. If this restoring force is sufficiently large, all the hunting motion can be eliminated.

(iv)   Providing adequate damping to the suspension is very useful in preventing primary hunting. However, it is almost ineffective in preventing secondary hunting; rather, excessive damping will also reduce the critical speed.

(v)   The critical speed of primary and secondary hunting becomes higher as wheel tread gradients are smaller, and it is almost proportional to the square root of the inverse of the tread gradient.

In addition, Matsudaira described the degree of front-back and left–right elastic coupling between axles and bogie frames as follows [4]:

There is an appropriate value for the axle support rigidity of the front, back, left, and right to maximize hunting critical speed, but in practice, it is desirable to have these values as high as possible, and it is better to set them at $k_{wx} = k_{wy} = 1500$–2000 kgf/mm or more. In addition, it is necessary to eliminate play that exists in axle support devices as much as possible. (Author's translation of the Japanese).

These findings were reflected in the prototype bogies for the test section, and running tests began in June 1962.

4.1.5 Criteria for Impact Lateral Force

At the commemorative lecture, Matsudaira discussed the derailment coefficient (= lateral force Q / wheel load P) (see Fig. 2.5). In the figure, the critical limit of Q/P is set at 1. However, even if Q/P exceeds 1, derailment will not occur if the time Q/P

Fig. 4.11 Time of impact lateral force and the limit of derailment coefficients *Source* Studies of the Tokaido Shinkansen, vol. 3, RTRI, 1962, p. 244

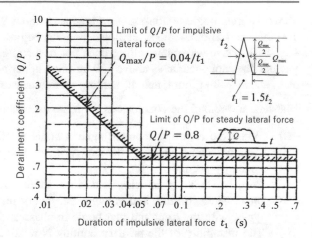

exceeds 1 is very short. Then, how should we decide the upper limit of time in excess of 1?

Matsudaira stated as follows [6]:

> There have been no definitive standards for the allowable limit of impact lateral force that acts on wheels at high speeds. Since Shinkansen vehicles run at very high speeds, large lateral force is expected to act on wheels and rails, so it is necessary to establish definitive standards for its tolerance before starting running tests. (Author's translation of the Japanese).

Using a roller rig, Matsudaira first made a model vehicle run at high speed, derailed it by violent hunting, and took high-speed photographs of the phenomenon to investigate the relationship with lateral force. Then he derived the relational expression among the collision speed of the wheel flange with the rail, the flange's jump height, the lateral force, and derailment. He revealed that the limiting value of the derailment coefficient due to impact lateral force is inversely proportional to impact time and proportional to the square root of flange height, and, furthermore, it becomes smaller when flange angles are smaller, when the ratio of vehicle's unsprung weight to sprung weight is smaller, and when the coefficient of friction between wheels and rails is larger.

Figure 4.11 shows the time of lateral impact force and the allowable derailment coefficient limit. The derailment coefficient limit is 0.8 when the duration is 50 ms or more, and $0.04/t_1$ when the time duration is 50 ms or less. The running tests on the test section were conducted while confirming safety according to these standards.

4.1.6 Running Test of Prototype Trains on the Test Section (1962–1964)

On June 26, 1962, running tests on the 32-km-long test section shown in Fig. 4.12 began. Test trains consisted of a two-car Train A and a four-car Train B. Train A was mainly used for power collection tests, and Train B was used for general tests. Since the power collection system's study was delayed compared to the hunting motion study, it might have required more running tests. All the technical issues, which were waiting for confirmation at actual speed, began testing at once. To evaluate hunting, lateral force and wheel loads on the axles at both ends of the train were measured constantly, and speeds were carefully increased while monitoring on a TV screen the movement of one wheelset for each car. The speed reached 200 km/h on October 31, 1962 and 256 km/h on March 30, 1963 (Photos 4.6, 4.7, 4.8, and 4.9).

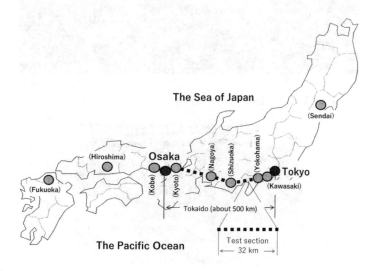

Fig. 4.12 Test section

Photo 4.6 Two-car Train A. (Provided by RTRI)

Photo 4.7　Four-car Train B. (Provided by RTRI)

Photo 4.8　Commemorative plate of a speed record of 256 km/h. (Author)

Photo 4.9　Scene in Train B. (Provided by RTRI)

4.1.6.1   Derailment Coefficient

The results of lateral force and wheel loads measurements were as follows:

(i)   Large lateral force or derailment coefficients appeared mostly at expansion/contraction joints, turnouts, and curved sections. The lateral force's acting time was relatively instantaneous, ranging from 1/10 to 1/40 s at joints and 1/4 to 1/20 s at curved sections.

(ii)   Fig. 4.13 shows the measured values of derailment coefficients. The values tend to increase gradually with speed. Still, they are almost within the safety limits except for those affected by severe hunting motion described next.

(iii)   From the results of these measurements, it was concluded that the test train was safe enough, at least up to about 250 km/h on a track as well maintained as this test section, if bogie hunting motion was completely eliminated.

(iv)   It was determined that when expansion/contraction joints are located in curves, special attention should be paid to the track's alignment and maintenance.

For the time being, prototype vehicles passed the test—with the proviso "if the bogie hunting is prevented." However, hunting had not yet been completely suppressed.

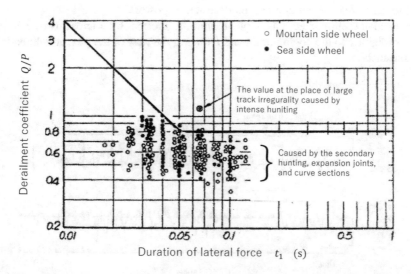

Fig. 4.13 Distribution of derailment coefficient *Source* Studies of the Tokaido Shinkansen, vol. 4, RTRI, 1963, p. 23

4.1.6.2 Hunting

During a series of tests, violent bogie hunting motion occurred in two bogies and frightened those on-board.

Figure 4.14 shows the waveform of the lateral force that occurred at 200 km/h.

Based on the time and distance scales and the wave number in the figure, the frequency of this hunting was 4.5–5.0 Hz and the wavelength was 11–12 m, which was much shorter than that of the geometrical hunting of this bogie (23.3 m) and that of the two-axle bogie (45.1 m). This kind of hunting had not been studied before. In other words, it was not caused by tread slopes but by flanges hitting rails and being pushed back.

As can be seen in the figure, the lateral force is shocking but not very instantaneous. It seems that the wheels are pushed back from the rails over a period of time. The lateral force reached a maximum of 7,500 kgf, and the derailment coefficient was 1.17. Although the coefficient exceeded the safety limit of 0.8, it was still in the margin of safety since its dangerous limit value, which was expected to cause an actual derailment, was 1.45. The hunting that occurred on the other bogie also occurred at a specific location at 235 km/h, with a hunting frequency of about 4.7 Hz and a wavelength of about 14.3 m. At this time, the lateral force and the lateral displacement were estimated to have reached 9,500 kgf and 10 mm, respectively, which left nine continuous bends in a 140 m track length with a maximum amplitude of 14 mm in peak to peak (thus demonstrating the well-founded fear of hunting and its destructive power).

Matsudaira eliminated this violent hunting motion by increasing the axle's support rigidity to the bogie frames and reducing the play of the pins at both ends of bolster anchors.

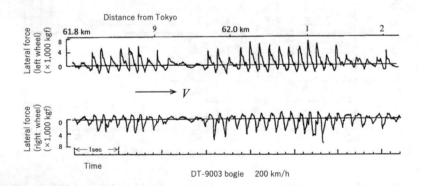

Fig. 4.14 Great lateral force on a wheel caused by hunting *Source* Studies of the Tokaido Shinkansen, vol. 4, RTRI, 1963, p. 258

4.1.7 Determining Bogie Specifications for Mass Production Vehicle

After the aforementioned process, Matsudaira decided the specifications of the Shinkansen bogie DT-200; specifications included the load on the center pivot, spring constant and damping constant of the primary and secondary suspension, spacing and friction coefficient of the side bearers, spacing of the bolster anchors, and so on (Photo 4.10).

The axle box suspension was specially devised to give the axles appropriate elasticity in both the front-rear and lateral directions by adopting a structure in which the bogie frame and the axle boxes are connected by metal plates with rubber bushes at both ends, as shown in Photo 4.11.

Photo 4.10 Type DT-200 bogie. (Provided by NIPPON SHARYO, LTD.)

Photo 4.11 Metal plate supporting the axle box. (Author)

The Shinkansen car equipped with this bogie showed excellent running stability during its 246 km/h trial run and kept a stable run during subsequent commercial operations.

Matsudaira's research on hunting motion began in 1947, two years after he came to the RTRI, and it culminated in this bogie 15 years later.

4.2 High-Speed Power Collection

The problem of Shinkansen's power collection system is roughly divided into two. One is catenary equipment and pantographs that keep contact even at high speeds, and the other is pantograph sliding strips that can withstand long-distance running. According to Fujii's analysis described in Sect. 1.5, the following four character-istics effectively realize high-speed overhead equipment-pantograph systems that keep contact at high speeds:

(i)   Increasing the average value K of the spring constant of catenary equipment.
(ii)   Reducing the nonuniform ratio ε of the spring constants of catenary equipment.
(iii) Reducing pantograph mass M.
(iv)   Giving pantograph a proper amount of damping.

4.2.1 Catenary Equipment for High-Speed Power Collection

In improving catenary equipment's performance, item(i) in the previous list is achieved by increasing the catenary equipment's total tension, and (ii) is achieved by devising its structure. Kumezawa,[1] who was the head of the power collection laboratory at the time, prioritized item (ii) and pursued catenary equipment with a uniform spring constant (i.e., catenary equipment with $\varepsilon = 0$).

Figure 4.15 shows the static uplift of various catenary equipment.

The figure shows that static uplifts are almost constant in three types: mesh-pattern catenary equipment, composite compound catenary equipment, and stitched compound catenary equipment. Kumezawa conducted on-track running tests using four types of catenary equipment: these three types and the one used on the Tokaido Line, which was standard compound catenary equipment.

4.2.1.1 Selection of Overhead Catenary Equipment
Figure 4.16 shows the configuration of these four types, and Photo. 4.12 shows KUMOYA 93,000, a dedicated car built in 1959 for the testing and inspection of catenary equipment. The car set a world record of 175 km/h of the narrow gauge railway on November 21, 1960. Incidentally, the "KU" in "KUMOYA" means cars with driver's cabs, "MO" means powered vehicles, and "YA" means noncommercial vehicles.

---

[1] Ikuro Kumezawa joined the Ministry of Railways in 1941. After serving the head of the power collection laboratory of RTRI, he later became a professor at the Tokyo University of Science.

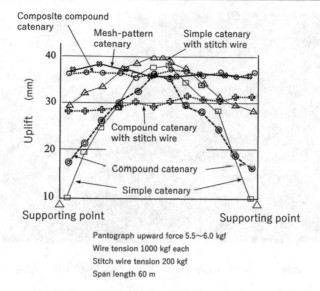

Fig. 4.15 Static uplift of catenary equipment. *Source* Kumezawa, I., "Overhead Catenary for High Speed", Railway Technical Research Report, no. 575, RTRI, 1967, p. 47

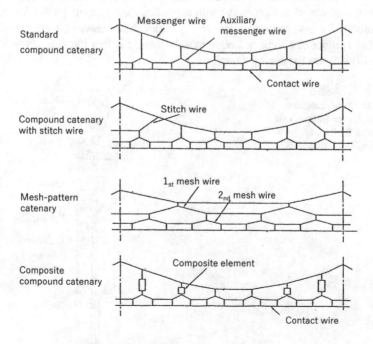

Fig. 4.16 Four types of catenary equipment

Photo 4.12 Catenary equipment inspection car KUMOYA 93,000. (Provided by RTRI)

In the stitched compound catenary and the mesh-pattern catenary equipment, the knotting of the wires that make up catenary equipment is ingenious. The composite compound type uses elements that have spring action (composite element) to even the spring stiffness of the catenary equipment between support points. In terms of the number of wires, the mesh-pattern type has four wires, and the others have three. The composite compound equipment's feature is that the composite element (Fig. 4.17) can have damping function.

Fig. 4.17 Composite element to absorb catenary vibration. *Source* Research on High-Speed Railways, Kenyusha Foundation, 1967, p. 416

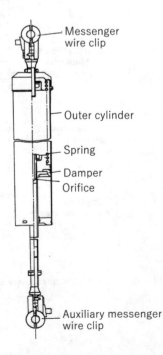

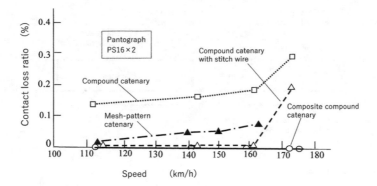

Fig. 4.18 Catenary equipment and contact loss characteristics. *Source* Studies of the Tokaido Shinkansen, vol. 2, RTRI, 1961, p. 27

Figure 4.18 shows the speed-to-contact-loss ratios up to 175 km/h for these four types of catenary equipment.

Commenting on these results, Kumezawa said as follows [7]:

> The movement of the pantographs and the composite compound catenary equipment showed that the pantographs pushed up the contact wire without vibration. After the pantograph passed, the wire dropped back to its original position without vibration, offering very stable power collection. In this way, the composite compound catenary dampens the wire's vibration quickly so it is less likely to cause resonance with pantographs than with other catenary equipment. (Author's translation of the Japanese).

The natural frequency of catenary equipment supported at 50 m intervals is about 1 Hz. Meanwhile, the Shinkansen trains at the time, which had pantographs at 50 m intervals, pushed up catenary equipment every second at a speed of 180 km/h, so it was necessary to suppress the catenary equipment's resonance that would occur at this speed. Kumezawa's preceding comment refers to this. After the aforementioned process, the composite compound catenary equipment was chosen for the Shinkansen in May 1961.

In 1954, large contact losses occurred at a speed of more than 100 km/h in a conventional line test. Seven years later, power collection at a speed of 200 km/h became realistic by using the new type of catenary equipment. The guiding principle was Fujii's analysis.

4.2.1.2   Tests on the Test Section

In the Shinkansen test section shown in Fig. 4.12, the standard compound catenary equipment was constructed along with the composite compound type (see Fig. 4.16) for performance comparison. Figure 4.19 and Photo 4.13 show two-car Train A used for the measurements.

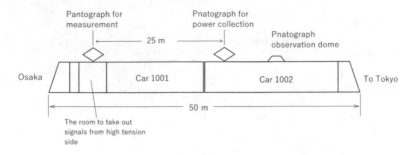

Fig. 4.19 Train A test car

Photo 4.13 Train A. (Provided by RTRI)

If a train has more than one pantograph, as does Train A, then the rear pantograph's condition would be worse than the front because the rear is subject to the catenary vibration caused by the front. Figure 4.20 shows the contact loss ratio in the rear pantograph of Train A, which had two pantographs. The contact loss of the composite compound equipment is smaller than that of the standard compound type. With the standard compound type, the contact loss increases sharply above 180 km/h, while that of the composite compound type remains small.

A 12-car Shinkansen train at that time consisted of six units, and each unit had two cars with eight motors and one pantograph. Thus, the train had six pantographs spaced 50 m apart, as shown in Fig. 4.21.

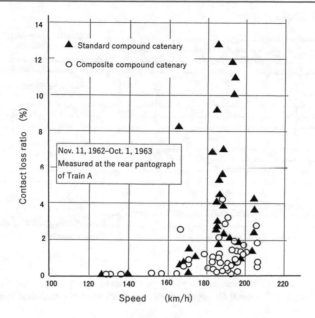

Fig. 4.20 Contact loss ratio of composite compound catenary equipment and standard compounded catenary equipment *Source* Research on High-speed Railways, Kenyusha Foundation, 1967, p. 419

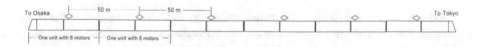

Fig. 4.21 Twelve-car train with six pantographs

The number and spacing of pantographs have a significant effect on power collection performance. Therefore, in the test section, tests were conducted to investigate the effect of multi-pantograph running. Since the number of tests was limited, the actual vibrations of catenary equipment and contact loss were not yet well understood before the opening. Hence, the catenary vibration by commercial vehicles was measured in March and July 1965 after the opening. Figure 4.22 shows the vibration waveforms of the two types of catenary equipment when six pantographs passed through.

Kumezawa commented on the waveforms as follows [7]:

It is natural that the uplift of the composite compound type at support points is almost twice as large as that of the standard compound type because the former eliminates the hard spot at support points and has a uniform spring constant across the span. Comparing the second and fourth waveforms from the top, the contact wire is pushed up obediently by the pantographs in the composite compound type. In the standard compound type, the contact

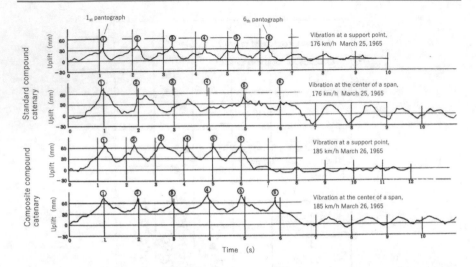

Fig. 4.22 Displacement and vibration of two types of catenary equipment by six pantographs. *Source* Kumezawa, I., "Overhead Catenary for High Speed", Railway Technical Research Report, no.575, RTRI, 1967, p. 87

wire movement becomes out of phase with the pantographs' movement around the third pantograph. From around the fourth to the sixth pantograph, it seems that the pantographs and the contact wire continue to collide, which is an alarming problem in the long run. (Author's translation of the Japanese).

Indeed, in the composite compound catenary, the vibrations are sufficiently damped that they do not interfere with the subsequent pantographs' movement. However, this large displacement of the composite compound catenary equipment, which Kumezawa mentioned, posed a serious problem combined with a unique section structure, as described next.

Generally, catenary equipment is spatially separated about every 1.5 km, and this separation space is called a section or an overlap. At a section, although being spatially separated, two contact wires are arranged horizontally parallel about 30 cm apart so that pantographs can travel smoothly, as shown in Fig. 4.23.

Regular sections consist of two sets of catenary equipment, as shown in Fig. 4.23, but the so-called twisted section consists of one set of catenary equipment; it was devised from the necessity of having two sections in one span. (See Sect. 4.9) Fig. 4.24 shows the structure of the twisted section. At point A in the figure, the three wires are arranged vertically. The arrangement is twisted from this state so that the auxiliary messenger wire becomes horizontally parallel to the contact wire at point B, forming a section with these two wires. As will be mentioned in Sect. 4.9, this section was a difficult and urgent measure.

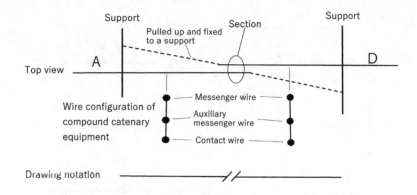

Fig. 4.23 Section structure of catenary equipment

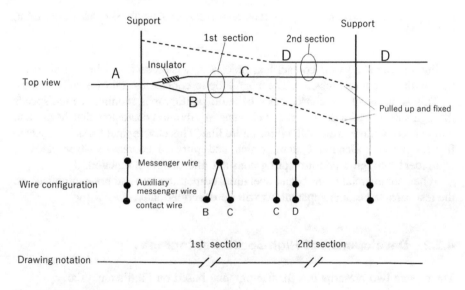

Fig. 4.24 Structure of the twisted section

The decision to adopt this type of section with a unique structure was made in November 1963. At that time, the basic tests of the power collection system, which had started in the summer of 1962, had already been completed, and it was less than a year before the business opening. At the end of November, construction to add twisted sections began, and at the same time, tests began to see how much contact losses would occur at this place. Naturally, large contact losses occurred, and the tests continued from December to January 1964 with repeated trial and error in an effort to reduce this problem.

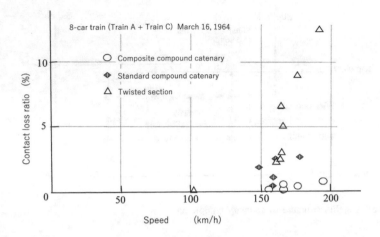

Fig. 4.25 Contact loss ratio at twisted section. *Source* Studies of the Tokaido Shinkansen, vol. 5, RTRI, 1964, p. 487

Figure 4.25 shows the contact loss ratio at a twisted section when an eight-car train with five pantographs ran. The magnitude of contact loss ratios is conspicuous.

This is an inevitable consequence of multi-pantographs running at high speeds through sections with discontinuous changes in vibration characteristics. More than 300 twisted sections were built on the entire line. This development was unexpected for the power collection system, which had pursued contact-loss-free catenary equipment through a uniform spring constant and damping function.

When commercial trains began operation, something that had gone unnoticed on the test section became apparent, as will be described later.

4.2.2 Development of High-Speed Pantographs

There were two prescriptions for pantographs based on Fujii's analysis:

- Reducing the equivalent mass of pantographs
- Adding appropriate damping to pantographs

In addition, high-speed pantographs had another challenge: making aerodynamics at high speeds harmless. The research process for these issues is described below.

Photo 4.14 shows the names of each part of a pantograph.

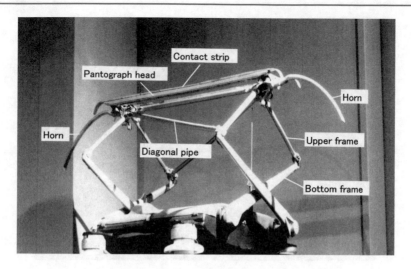

Photo 4.14 Names of pantograph parts. (Author)

4.2.2.1 Reduction of Pantograph Equivalent Mass

The equivalent mass M is the mass of all the parts that move with the pantograph head's movement converted to the pantograph head position. It is measured as shown in Fig. 4.26.

That is, with the pantograph raised, the pantograph head is pulled down by a spring with a spring constant K and is freely vibrated up and down to read the vibration period T. The equivalent mass M is obtained from the following relationship:

$$T = 2\pi\sqrt{M/K}$$

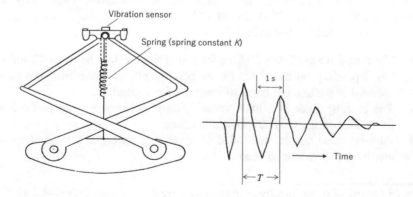

Fig. 4.26 Measurement of equivalent mass

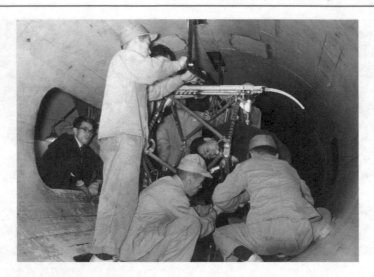

Photo 4.15 Pantograph wind-tunnel test. (Provided by RTRI)

To reduce M, the pantograph head must first be lightened, but since sliding strips have a certain weight, there is limitation to the head's weight reduction.

Next, frameworks can be made smaller if the variation of contact wires' height is small. As a result, the equivalent mass of the Shinkansen pantograph, named PS-200, became about 12 kg, much smaller than that of conventional pantographs (20–35 kg).

4.2.2.2 Control of Aerodynamic Force

It is necessary to put an actual pantograph in a wind tunnel and send wind at the train speed to determine what kind of force is generated and how the pantograph moves in a high-speed air stream. The pantograph development team led by Arimoto[2] conducted wind-tunnel tests three times: November 1959, July and November 1960, and June 1961 (Photo 4.15).

The tests revealed the following:

(i)   The drag force at 60 m/s (216 km/h) varied significantly between 40 and 110 kgf depending on the shape of the pantograph, so windshield covers are required at pantograph bases to keep the force constant.

(ii)   The push-up force at 60 m/s varied greatly, between −6 kgf and + 3 kgf depending on the pantograph head's shape.

(iii) Push-up force can be controlled by changing pantograph head shapes or attaching protrusions to heads.

---

[2] Hiroshi Arimoto joined the Ministry of Transportation and Communications in 1944. He became a senior researcher at RTRI in 1950, and later became head of the power collection laboratory of RTRI.

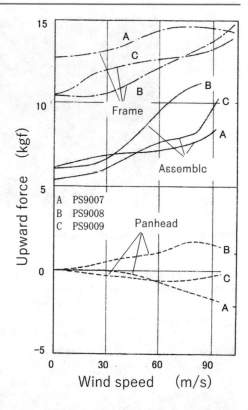

Fig. 4.27 Vertical aerodynamic force of prototype pantographs. *Source* Research on High-speed Railways, Kenyusha Foundation, 1967, p. 431

(iv) Since the push-up force generated from frameworks is about 1–4 kgf, it is important to design pantograph heads to counteract this force.

Three prototypes, PS9007, PS9008, and PS9009, were designed to achieve push-up force of 5.5 kgf at a standstill and 7.5 kgf at 200 km/h. Figure 4.27 shows the prototypes' resultant upward force and its components (vertical aerodynamic force of the frameworks and pantograph head).

In February 1963, Arimoto and his colleagues installed the prototype pantographs on Train A and conducted the first high-speed running tests. The results showed that aerodynamics was controlled almost exactly as expected, and the pantographs had no unstable vibrations even at high speeds. The push-up force at 200 km/h was slightly greater than that at a standstill, as designed.

In October 1963, Arimoto's team tested the final prototype and fixed the specifications for the commercial vehicle's pantograph. Photo 4.16 shows the complete model, named PS200.

Table 4.2 shows the mechanical specifications of the power collection system at the time of opening.

Wave propagation speed shown in the table is the traveling speed of transverse waves along wires. Fujii proposed it in 1959 as the theoretical speed limitation of catenary equipment. Since then, this speed has become the most significant figure of

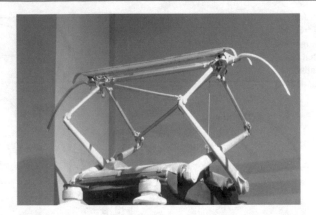

Photo 4.16 PS200 pantograph. (Author)

Table 4.2 Mechanical specifications of the power collection system at the time of opening. *Source* Shimomae, T. Manabe, K. Aboshi, M., How Was the Continuous Arc of the Shinkansen Pantograph Eliminated?, Japan Railway Electric Engineering Association, 2008, p. 16

Catenary Equipment: Composite Compound Catenary		Pantograph: Type PS200	
Span length	60 m	Equivqlent mass	12.7 kg
Wire tension	29.4 kN (3,000 kgf) in total	Panhead mass	5 kg
Composite element	Spring constant 490–1,960 N/m	Static upward force	54 N
Wave propagation speed • Catenary equipment • Contact wire	412 km/h 358 km/h	Aerodynamic upward force	20 N at 200 km/h
Mean value of spring constants	1,570 N/m	Damper	81 Ns/m: only for downward
Non-uniformity ratio of spring constants, ε	0	Vertical working range	500 mm

power collection systems. In the 1980s, Manabe[3] and Aboshi[4] studied the contact force variation between contact wires and pantographs and found that the wave propagation speed of the contact wire is more important than that of entire catenary equipment. The catenary equipment in use today follows their recommendation that the contact wires' wave propagation speed should be at least 140% of train speed.

---

[3] Katsushi Manabe joined JNR in 1967. He later became head of the power collection laboratory of RTRI after served as a senior researcher.

[4] Mitsuo Aboshi joined JNR in 1979. He later became head of the power collection laboratory of RTRI after served as a senior researcher.

Wave propagation speed v_p, wire density ρ, and wire tension T are related to the following equation:

$$v_p = \sqrt{T/\rho}$$

Judging from the broken pantographs of the Société Nationale des Chemins de fer Français (SNCF) locomotives that set a speed record of 331 km/h in 1955, it is assumed that the catenary equipment's wave propagation speed was too low for the train speed at the time.

4.2.3   Development of Pantograph Contact Strip for the Shinkansen

Material wear is a complex phenomenon influenced by many factors, including lubrication, contact pressure, sliding speed, material itself, and mating material. In addition, pantograph contact strips' wear is also affected by the electric currents that flow between the contacting surfaces. The exact wear mechanism in this case is unknown. As a result, the development of the sliding strips for the Shinkansen required a series of trial-and-error tests. Masaru Iwase, who was in charge of the development of the contact strips, wrote in his memoir [8]:

> In 1955, when SNCF set a speed record of 331 km/h, the power supply voltage was 1,700 V at the time of the test, the current that flowed to the pantograph was about 4,500 A, and the contact loss ratio was about 25%. This large contact loss ratio and burned-out pantographs gave a sense of how difficult it would be to collect power at high speeds. (Author's translation of the Japanese).

Therefore, it was determined that the Shinkansen's contact strips must withstand at least seven to eight round trips between Tokyo and Osaka, or seven to eight thousand kilometers, even if they experienced a considerable amount of contact loss.

Iwase tested 31 contact strips using the test equipment shown in Fig. 4.28 and selected the following three:

- Copper-based sintered material (NT-1)
- Copper-based casting material (NT-2)
- Iron-based sintered material (NT-3).

These were installed on the test trains in 1962 after trials on AC-electrified conventional lines.

However, these strips' relative superiority was not clear because of the small distance traveled on the test section. In the end, only NT-3 (iron-based sintered alloy) met the target in the pre-opening runs of commercial trains, so it was decided that this would be the contact strips for the Shinkansen trains.

In 1968, newly developed copper-based sintered contact strips were used together with NT-3, and the life of the strips increased dramatically to between

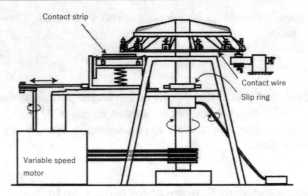

Fig. 4.28 Wear tester for pantograph contact strip. *Source* Iwase, M., "Sliding Contact Abrasion between Overhead Wire and Collector, and Effect of Wear upon Power Collection (1)", Railway Technical Research Report, no. 53, RTRI, 1959, p. 7.

15,000 and 17,000 km. After that, as train speed increased and the number of pantographs was reduced, copper-based products disappeared, and currently only iron-based products with enhanced wear performance are used.

4.2.4 Post-opening Problems

Two and a half months after the business opening, the first significant failure occurred. An overhead wire broke at a twisted section, and damaged pantographs ran for 3 km, while destroying the catenary equipment. Due to this failure, trains did not resume operation for 14 hours. Table 4.3 shows the number of wire breaks in the composite compound catenary equipment in the 20 years from the opening to 1984.

These breaks were caused by detachment of overhead wire fittings, rapid wear of contact wires at twisted sections, fatigue failure of contact wires above track junctions, and so on.

Table 4.3 Number of wire breaks in the composite compound catenary equipment

	Number of wire breaks in 20 years[a]
Messenger wire	10
Auxiliary messenger wire	4
Contact wire	11
Total	25

[a] 1964–1984

Arimoto, who developed the heavy compound catenary equipment that modified the weaknesses of the composite compound type, said [9]:

> The composite compound catenary equipment was called the "ultimate catenary equipment" and used the composite elements to eliminate the inequality of spring constants and to dampen wire vibrations. That is, it intended to be the one with no defects. However, when it was used in practice, considerable problems developed. (Author's translation of the Japanese).

In other words, large flexibility of the composite compound catenary equipment with uniform spring constant led to large displacement of contact wires, which resulted in loose overhead fittings and wire fatigue due to numerous pantograph runs. Furthermore, the overhead wires were lifted significantly by strong wind, which made it easier for pantographs to hit overhead wire fittings. In addition, at the twisted sections, contact wires and pantographs were prone to collide, which caused rapid wires' wear. It was not possible to predict the system's vulnerability during the tests on the test section.

The weakness of the catenary equipment that emphasized only the spring constant's uniformity was modified by adopting another type of catenary equipment called the heavy compound catenary equipment. This type of catenary equipment, while allowing a small but nonzero value of spring constant inequality, focused on increasing the average value K of spring constants. Total tension was increased from 3,000 to 5,500 kgf, and the composite elements were eliminated.

On the Sanyo Shinkansen, the feedeing system was also changed from the BT feeding system to the AT feeding system (see Sect. 4.9), the twisted sections were also eliminated (see Sect. 4.9), and the number of pantographs was reduced by connecting pantographs with a cable. Since then, failure of the power collection system has disappeared.

By 1989, all the composite compound catenary equipment was converted to the heavy compound catenary type. The composite compound catenary equipment was used for 25 years and required a great deal of maintenance work.

Although there was a problem with stability under the BT feeding system and multiple-pantograph trains, the composite compound catenary equipment and PS200 pantographs are significant in the high-speed railway's history in that they made the first power collection system capable of exceeding 200 km/h.

Catenary equipment vibrates due to pantographs' passage. The vibration propagates back and forth while partially reflected by support points. Subsequent pantographs generate new vibrations under the vibrating catenary's influence, so the vibrations become more complicated. Pursuing power collection systems that do not lose contact even at high speeds requires knowing the contact force between contact wires and the pantographs under such vibrations. However, it is not realistic to find general analytical solutions for this phenomenon. Fujii's analysis clarified the cause of contact losses that occurred at every catenary support, the first barrier to high-speed power collection.

In 1969, to simulate more closely the systems' dynamics, Noburo Ehara[5] developed a computer-based digital simulation method, which was improved at RTRI and became a powerful tool for the development of high-speed power collection systems. Later, an analysis of the fluctuation of contact force in the hanger cycle of catenary equipment[6, 7] which becomes apparent with increase in speed, and reduction in the number of pantographs per train led to the current stable power collection systems at speeds over 300 km/h.

4.3 Shinkansen Track

As Hoshino said in the commemorative lecture, the problem of high-speed railway tracks is durability. Kentaro Matsubara,[8] who decided the track specifications for the Shinkansen, stated in his book [10]:

> As for Shinkansen track structure, there was a concern that ballast outflows would be large and maintenance could not keep up with in case of the conventional track structure, so asphalt track bed or non-track bed structures were considered, and several prototypes were tested. However, after that, it turned out that the maintenance amount could be kept to an acceptable level by improving the design of rail fastening devices, rails, sleepers, and track beds, even if the conventional track structure was adopted at high speeds of 200 km/h. (Author's translation of the Japanese).

4.3.1 Decision on Track Structure

4.3.1.1 Track Deterioration Theory

Since tracks are designed to be used while repairing the damage caused by trains, the challenge was the cost effectiveness of construction and maintenance costs combined.

Once a track structure is determined, the construction cost can be calculated, but the maintenance cost depends on the conditions of use. Based on the track dynamics described in Sect. 1.3, Sato developed a track deterioration theory that expresses the amount of track deterioration as a function of train weight, speed, number of transits, and track structure. Sato clarified that the conventional track structure could be applied to the Shinkansen, which would operate in an unexperienced speed range, if appropriate design decisions were made.

---

[5] Noburo Ehara was a graduate student at the University of Tokyo at the time and later became a professor at Meiji University.

[6] Manabe, K., "High-speed performance of overhead catenary equipment and pantograph system", *Transactions of the Japan Society of Mechanical Engineers* (JSME), C, no. 505 and 512, 1988 and 1989.

[7] Aboshi, M., Manabe, K., "Analysis of contact force fluctuation in catenary equipment and pantograph system", *RTRI Research Report*, no. 7, RTRI, 1999.

[8] Kentaro Matsubara joined the Ministry of Railways in 1941. He was manager of the Shinkansen track division in the head office when he decided the Shinkansen track specifications. He later became Deputy Chief Engineer of JNR and then Deputy Director of RTRI.

The track deterioration theory was constructed as follows:

(i) Define track deterioration factor Δ and express it as the product of load factor L, structure factor M, and state factor N.

$$\Delta = \text{load factor } L \times \text{ structure factor } M \times \text{ state factor } N$$

(ii) The load factor L expresses the effect of the force exerted by trains and is the product of vehicle factor K, tonnage T, and speed V.

$$L = \text{vehicle factor } K \times \text{ tonnage } T \times \text{ speed } V$$

Based on the track dynamics described in Sect. 1.3 and experiments conducted, Sato derived the relationship between the structure factor M and the rail stiffness, sleeper spacing, ballast thickness, rail pad elasticity, and so on.

Table 4.4 shows the ratio of the estimated structure factor of the first-class track of Deutsche Bundesbahn (hereafter referred to as DB) to that when the DB's rail fastener spring constant of 300 tf/cm (1tf = 1,000 kgf = 9,800 N) and ballast thickness of 35 cm are changed to 100 tf/cm and 30 cm, respectively. The structure factor is reduced by about 22% (from 1.28 to 1) due to these changes (track deterioration is reduced).

Table 4.5 shows the estimated load factors of the Shinkansen and DB's first-class line at that time.

According to Table 4.5, the load factor for the Shinkansen was expected to be 2.4 to 3.0 times that of DB's first-class lines. This means that if their maintenance are performed in the same way, the Shinkansen would have to build a track that is 2.4 to 3 times as strong as the DB's.

The smaller the M, the less track deterioration, but the higher the construction cost, so a track with less tonnage does not need to be built to withstand extreme conditions; M should be determined to minimize the total of capital and maintenance costs.

Table 4.4 Estimation of structure factor. *Source* Research on High-Speed Railways, Kenyusha Foundation, 1967, p. 81

		Unit	German track of first-class line	New track
Track structure	Rail	kg/m	49	50
	Sleeper space	cm	60	63
	Spring constant of rail pad	tf/cm[a]	300	100
	Thickness of track bed	cm	35	30
Ratio of M (Track structure factor)			1.28	1

[a] 1 tf = 1,000kgf

Table 4.5 Estimation of load factor. *Source* Research on High-Speed Railways, Kenyusha Foundation, 1967, p. 80

		Vehicle Factor K	Tonnage/Year T ($\times$10,000 tons)			Speed V (km/h)	Load factor $L =$ KTV		
			1969	1972	1975		1969	1972	1975
Shinkansen	Passenger train	0.18	2,200	2,400	2,600	148	2.40	2.71	3.08
	Freight train	0.22	800	1,000	1,300	100			
German	Passenger train	0.24		650		85	1.00		
	Freight train	0.28		1,050		63			

4.3.1.2 Desired Track Structure

Taking the Shinkansen's expected tonnage, construction costs, and maintenance fees into consideration, Sato determined that it would be most economical to double the track strength of DB's track and extend the interval between major repairs. Table 4.6 shows the track structure factors that minimize the total cost based on the assumed tonnage.

After the aforementioned process, Matsubara decided the specifications for the Shinkansen track. Table 4.7 shows a comparison between the Shinkansen track and the existing track used on the Tokaido Line.

In his book, Matsubara describes the design policy as follows: [10]

The Shinkansen track was designed to be 1.61 (= 1/0.62) times stronger than that of the conventional line so that the degree of track deterioration would be almost the same. (Author's translation of the Japanese).

At the memorial lecture, Hoshino said, "The current track structure is expected to be completely unsuitable for super high-speed trains running at 2 to 2.5 times the current speed."

However, the track deterioration theory based on track dynamics made it clear that conventional structure could be applied, which was one of the factors that made it possible to open the Shinkansen in 1964.

Table 4.6 Structure coefficients to minimize the total cost. *Source* Research on High-Speed Railways, Kenyusha Foundation, 1967, p. 84

	1969	1972	1975
Tonnage/year (10,000 tons)	3,000	3,400	3,900
Track structure factor M^a	0.66	0.62	0.58
Rail (kg/m)	52	54	56
Large-scale repair interval (year)	1.5	1.5	1
Replacement interval (year)	15–20	15–20	15–20

[a] Ratio with German track as 1.0

Table 4.7 Comparison of the Shinkansen track and the Tokaido Line track. *Source* Matsubara, K., Shinkansen Track, Japan Railway Civil Engineering Association, 1969, p. 27

		Shinkansen	Tokaido Line
Load factor L	Ratio of vehicle factor	0.77	1
	Ratio of tonnage	1	1
	Ratio of speed	2	1
	Ratio of load factor L	1.54	1
Track structure factor M	Rail (kg/m)	53.3	50.4
	Sleeper space (cm)	58	58
	Spring constant of rail pad (tf/cm) [a]	90	100
	Trackbed thickness (cm)	30	25
	Ratio of track structure factor M	0.62	1
Ratio of track deterioration index[b] $L \times M$		0.95	1

[a] 1 tf = 1,000 kgf
[b] State factor N is 1.0 for both

The basic research that had been conducted before the Shinkansen project started played an essential role in the field of track design.

4.3.2 Measurement of Vibration and Stress of the Track on the Test Section

On the test section, vibration acceleration, displacement, and stress were measured at various parts of the track, including rails, track beds, and sleepers.

The results were as follows:

(i) Figure 4.29 shows the rail acceleration amplitude. As expected, it increased almost in proportion to the speed. The variation in the measurements was due to the proximity or nonproximity of rail welds.

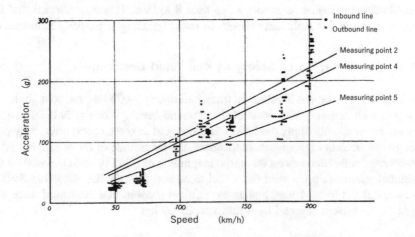

Fig. 4.29 Rail acceleration. *Source* Research on High-Speed Railways, Kenyusha Foundation, 1967, p. 86

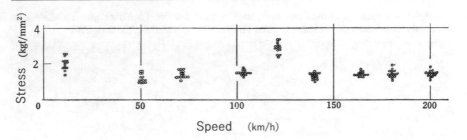

Fig. 4.30 Rail stress. *Source* Sato, Y., Track Dynamics, Tetsudo Gengyousha, 1964, p. 30

(ii)   The decay rate of acceleration was about 1/2 between the sleeper and the track bed and about 1/40–1/60 between the rail and the sleeper.
(iii)   As expected, the rail stress was independent of speed as shown in Fig. 4.30 and was less than 4 kgf/mm^2, which was not a problem.
(iv)   Expansion/contraction joints and branching devices were not a problem when they were carefully assembled and welded.

No particular problems were found in high-speed range, and track's function was confirmed almost exactly as designed.

4.3.3 Track Buckling Test

In April 1964, a test was conducted to confirm the buckling strength of the Shinkansen track. Specifically, this was a test in which the rails were heated to cause buckling on a track that had initial irregularities and lateral ballast resistance reduced to 4 kgf/cm. When heated to 50–85 °C by mobile acetylene burners, rail axial force reached 80–90 tf but buckling did not occur.

Meanwhile, since the Shinkansen track is laid in such a way that the maximum axial force does not exceed 65 tf even at the maximum temperature, and since the lateral ballast resistance is usually more than 8 kgf/cm, it was confirmed that the Shinkansen track has sufficient strength to resist buckling if properly maintained.

4.3.4 Confirmation of Safety at Rail Weld Breakage

The long rails laid on the Shinkansen track had nearly 80,000 welds, so it could not be stated with certainty that none of them would break. When rails break, signal systems automatically apply the brakes to trains, but in some cases, trains may pass through the break at high speeds. In this case, the magnitude of the misalignment of the separated rails becomes an issue (the magnitude of q in Fig. 4.31). Based on the positional relationship between the wheel tread and the rail, the allowable limit of q was calculated to be 4 mm, and an experiment showed that the lateral force that displaces the broken rail end by 4 mm was 2,000 kgf.

Fig. 4.31 Wheel at a rail break. *Source* Research on High-Speed Railways, Kenyusha Foundation, 1967, p. 94

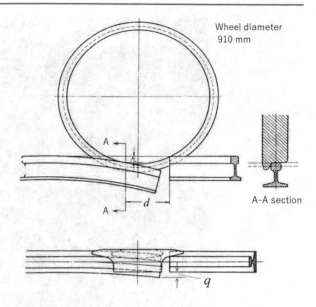

Preliminary tests were conducted at RTRI using a 1/5 scale model bogie and a roller rig with circumference of 10 m. The equipment was typically used for investigating the relationship between track irregularity and vehicle vibration. The tests showed that the model bogie did not derail at speeds equivalent to 80 to 100 km/h in real vehicles under the condition where lateral misalignments equivalent to 5 mm and 8 mm in actual rails were given to one rail of the roller, and lateral forces equivalent to 1,000 kgf and 6,000 kgf in a real bogie were given to the front axle (Photo 4.17).

Running tests on the test track were conducted in March 1964 by cutting one rail to make a gap of 20 mm. To prevent derailment during the tests, a 10-m guard rail was placed along the gap. Tests were repeated at speeds from 60 to 200 km/h to measure rail displacement and rail stress at the cut point.

Figure 4.32 shows the lateral rail displacement and lateral force that occurred when a train ran through the cut point. The figure shows that even if 3σ was added to the average value, the lateral rail displacement was about 1.4 mm, and the lateral force was about 1,000 kgf. Since the stresses of the rail and rail fastening devices were acceptable, it was concluded that trains could run safely at 200 km/h at a weld break.

In addition to the above, the following issues were studied:

- Determining the cross-sectional shape and material of the rail
- Developing rail elastic fasteners and concrete sleepers
- Developing turnouts and expansion joints
- Developing rail welding methods and nondestructive inspection methods for welded parts
- Developing high-speed track inspection vehicles.

Photo 4.17 Preliminary test. (Provided by RTRI)

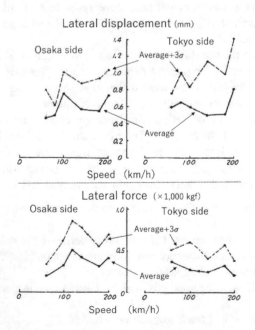

Fig. 4.32 Rail displacement and lateral force at a cut point. *Source* Studies of the Tokaido Shinkansen, vol. 5, RTRI, 1964, p. 56

Photo 4.18 Track deterioration tester. (Provided by RTRI)

4.4   Strength of Vehicle

As described in Sect. 1.2, after the war, there were many technological advances in railcars, such as body strength, weight reduction, and aerodynamic resistance. These were the basis for realizing the Shinkansen vehicle.

The typical Japanese narrow gauge car body was 19.5 m long and weighed 7–9 tons. Since the Shinkansen vehicle body had to be 1.25 times as long and 1.2 times as wide (that is, 1.5 times the volume) and weighs 8 tons at most while retaining sufficient strength and rigidity, the design was challenging. After designs and strength checks by calculation were repeated several times, in December 1960, a car body with a frame length of 24.5 m named Prototype 0 was finally completed and passed static load tests. The design method for high-speed railcars was established at this stage, and six test cars were produced for the Shinkansen test section.

4.4.1   Airtight Body

One of the new findings related to vehicle bodies in the running tests on the test section was the so-called ear-popping phenomenon that occurred when trains entered tunnels (see Sect. 4.7). Running tests were conducted with one of the test cars rigorously sealed up. Reduced ear-popping was confirmed, but airtight car bodies became subjected to repeated load fluctuations due to air pressure. There had never been pressure-resistance problems in railcars, so making railcars airtight significantly affected car body structure.

A strength test against atmospheric pressure change was carried out by setting the car's outside pressure at –500 mmAq. Since –500 mmAq is a negative pressure of 500 kgf per square meter, one side of the car, which was 2.3 m high and 25 m long, would be subjected to a negative pressure of over 28,000 kgf. The test was conducted in March 1963 by placing a large plastic bag in a car and pressurizing it with air to 500 mmAq. As a result, it was found that there were strength problems in welding parts, and the countermeasures were reflected in mass production cars.

Window glass was also strengthened to match the airtightness and was confirmed to be strong enough when two trains passed each other in a tunnel.

4.4.2 Static Load Tests

Static load tests (vertical load, longitudinal compression, torsion, bending, and torsional natural frequency) for a mass-produced car were conducted in October 1963 to confirm the prescribed strength and rigidity requirements (i.e., with twice as many passengers as the capacity onboard, the vehicle could run safely under a vertical vibration acceleration of 0.1 g). It was also confirmed that the vehicle did not deform plastically on the frame when it received compression force of 100 tf from the coupler, and the vehicle did not leave wrinkles on the outer plate even if unevenly supported during transportation from manufacturers to railcar bases or during inspection at depots (even supported at three points due to the floating of one support point).

Figure 4.33 shows the bending under vertical loading and horizontal compression, and Photo 4.19 shows a static load test.

4.4.3 Axle Strength

When considering railways' safety, the importance of axles is indisputable. The first study of axle strength in Japan began in the early 1930s and investigated the fatigue strength of wheel mating places. In 1949, induction hardening was tested on traction motor shafts before its application to vehicle axles, which began the following year. High-frequency induction hardening is a metal hardening method that produces surface hardening and compressive residual stress on axles by utilizing the electromagnetic characteristics that high-frequency electric currents gather on the conductor's surfaces (skin effect). Compressive residual stress is said to improve fatigue strength because it closes the wounds created by tension.

In 1954, as described in Sect. 1.2, it became possible to measure stress on rotating axles using strain gauges via slip ring devices. Axle strength measurements were made during the high-speed tests of Odakyu SE train (in 1957), Kodama trains (1959), and KUMOYA 93,000 catenary equipment inspection car (1960); these measurements led to rational designs based on the data.

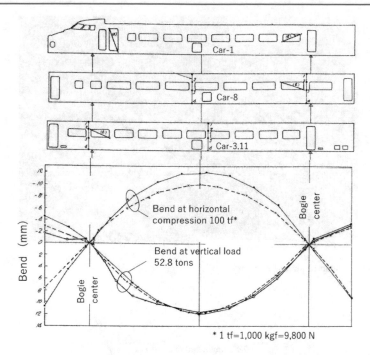

Fig. 4.33 Bending of car body under static load. *Source* Research on High-Speed Railways, Kenyusha Foundation, 1967, p. 187

Figure 4.34 shows lateral force acting on wheels and axle bending stress. It indicates that axle stress increases roughly in proportion to lateral force.

Figure 4.35 shows the relationship between speed and wheel load fluctuation, where $P_d$max, $P_d$min, $\overline{P}$, and σ are the maximum dynamic load, the minimum dynamic load, the mean of the loads, and the standard deviation of the fluctuation, respectively. This figure indicates that when the speed reaches 200 km/h, the dynamic wheel load would become about 40–160% of the static load.

Based on these data, it was assumed that wheel loads at 200 km/h would be 1.7 times the stationary wheel load, lateral force would be 50% of the static wheel load, and the maximum stress on axles would be 6.6 kgf/mm^2.

In addition, based on the stress and scratches of railcar axles in the Tokyo suburbs, the desirable maximum stress with the current axle material was considered to be about 7 kgf/mm^2. Since the estimated maximum stress of 6.6 kgf/mm^2 was within this range and the effect of induction hardening on axles would be significant, it was expected that the Shinkansen axles would have sufficient strength.

Photo 4.19 Scene of static load test[9]

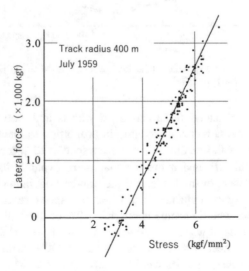

Fig. 4.34 Wheel lateral force and axle bending stress. *Source* Studies of the Tokaido Shinkansen, vol. 1, RTRI, 1960, p. 81

Figure 4.36 shows the frequency of axle bending stress when a test vehicle was traveling at 200 km/h on the test section. The horizontal axis is the ratio of dynamic stress σ_d due to wheel loads and lateral force to the stress under static load σ_{st}. The figure shows that axles have a sufficient margin in strength.

---

[9] Research on High-Speed Railways, RTRI, Ed., Kenyusha Foundation, 1967, p. 173.

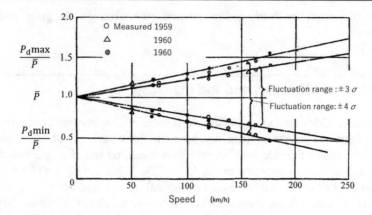

Fig. 4.35 Speed versus wheel load variation. *Source* Studies of the Tokaido Shinkansen, vol. 2, RTRI, 1961, p. 10

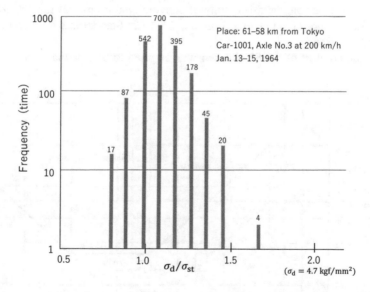

Fig. 4.36 Axle bending stress on the test section. *Source* Studies of the Tokaido Shinkansen, vol. 5, RTRI, 1964, p. 274

Regarding the detection of scratches on axles, in 1950, ultrasonic vertical flaw detection made it possible to detect scratches whose length ranged from 2 to 3 mm. In 1958, ultrasonic oblique flaw detection made it possible to detect scratches from 0.5 to 1 mm long. When the Shinkansen opened in 1964, it was possible to detect scratches of about 0.1 mm long by using the magnetic particle inspection method.

Tests conducted also confirmed the strength of bogies, wheels, gears, and other elements.

4.5 Vehicle Vibration and Ride Quality

In September 1958, sensory evaluation of ride comfort was conducted three times to investigate the effects of train acceleration, deceleration, and curve passage. Figure 4.37 shows the ride comfort standards based on the aforementioned evaluations and research conducted in the USA. The solid, dotted, and broken lines represent the ride comfort level related to vertical, lateral, and front-back vibration, respectively. Although the three lines have different acceleration values, they have the same ride comfort, and this is defined as the ride quality index of 1. For example, when the vibration at 10 Hz is 0.066 g vertically, 0 0.04 g laterally, and 0.05 g back and forth, these three vibrations will result in the same ride quality index of 2.

Using this criterion, the ride quality of conventional trains was rated as follows:

1 or Less: Very Good, 1 to 1.5: Good, 1.5 to 2: Standard, 2 to 3: Bad, 3 and Above: Very Bad.

The ride comfort of the Shinkansen was targeted to be less than 1.5.

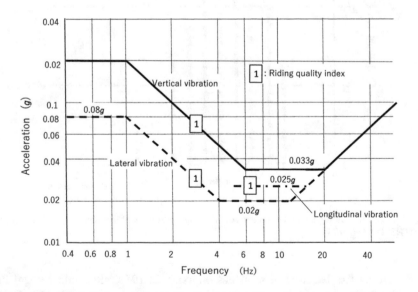

Fig. 4.37 Ride quality standards. *Source* Studies of the Tokaido Shinkansen, vol. 4, RTRI, 1963, p. 244

4.5.1 Development of Air Spring

To reduce vibration being transmitted to vehicle bodies, the spring constants of bogie springs should be reduced. However, for a train with many vehicles connected, it is necessary to keep the floor height from the rails of full cars and empty cars within a certain limit. It is difficult to achieve this goal using metal springs, but with air springs, spring constants can be drastically reduced because the height of the car floors can be kept constant regardless of load conditions via height adjustment devices. This is why air springs are suitable for railway vehicles.

In his memoirs, Matsudaira described the development of air springs for railcars as follows [11]:

> Research on air springs for railway vehicles began at the Railway Technical Research Institute in 1955. At the time, I learned from a magazine article that air springs for automobiles had already been developed in the United States and were used in Greyhound buses, but their technical details were unknown.
>
> After much research and many trials, we succeeded in developing air springs for railway cars in 1958 and adopted them for the "Asakaze" sleeper train between Tokyo and Hakata and the "Kodama" limited express train between Tokyo and Osaka, with good results.
>
> Due to this success, air springs were adopted for all later high-class trains, and they became one of the major features of Japanese railcars.
>
> The air spring of this period was the three-stage bellows type, which replaced conventional coil springs as the secondary suspension.
>
> Sufficient research was conducted at that time on the vertical behavior of this type of air spring, and in addition to the static spring constant, the dynamic spring constant and damping coefficient, which included the effect of the throttle installed in the air passage between spring bodies and auxiliary tanks, were accurately calculated and effectively used in the design of bogie suspension systems.
>
> … In bogies with air springs at that time, conventional swing hangar link systems were still used to alleviate left and right vibration. However, the next step was to discontinue this linkage system and use air springs to produce lateral spring action.
>
> Beginning around 1958, a prototype bogie of this type was made, and at the same time, research on the lateral behavior of air springs became more active. (Author's translation of the Japanese).

Kunieda, who was involved in the development of air springs, described the situation at the time as follows [12]:

> Although trains with air springs appeared one after another, almost all of the research on air springs was experimental. There was no guideline for railcars with air springs, so all were produced first and then improved experimentally. However, even though experiments were repeated many times, the extent to which they could reveal the spring characteristics was limited, and in many cases, air springs have not been fully exploited. (Author's translation of the Japanese).
>
> Kunieda's analysis revealed the air spring characteristics and led to the best design for the use of air springs.

4.5.1.1 Vertical Vibration Characteristics of Air Spring

Figure 4.38 shows a three-stage bellows-type air spring's analytical model, where p, V, and w are the internal pressure, the bellows volume, and the specific gravity of air, respectively, and p_t, V_t, w_t are the respective values of the reservoir. When external force P is applied, the bellows displace by x, and the air in the bellows flows through the orifice into the reservoir, creating resistance depending on the airflow speed. In other words, the orifice provides damping function to the air spring. Since the airflow depends on the amplitude and frequency of x, the spring constant and damping factor also depend on these. Therefore, depending on many factors, such as the amplitude and frequency of x, volume of bellows and reservoirs, size of orifices, and weight of vehicle bodies, vibration isolation characteristics will vary in a complex way.

Based on this analysis, Kunieda expressed the characteristics of the air spring as follows:

$$P = P_0 + Kx + C\ddot{x}$$

where P_0, K, C are the static load, spring constant, and damping coefficient, respectively.

The preceding equation is simple, but the coefficients K and C are all complex polynomials that include the amplitude and frequency of x.

Figure 4.39 shows the static characteristics of the air spring used in the "Asa-kaze" sleeper train (Asakaze means a morning breeze). It can be seen from the figure that calculated and experimental values are in good agreement and that spring constants are kept constant by the reservoir.

Kunieda stated that [13]:

> Through theoretical analysis, the dynamic spring constant, damping coefficient, and throttle's optimum dimensions have been clarified. The results of the analysis have already been applied to the air spring design for the "Kodama" and "Asakaze," and many on-board tests have shown that the results are as designed, so the analysis can serve as a guide to the Shinkansen car design. (Author's translation of the Japanese).

Fig. 4.38 Analytical model of bellows-type air spring. *Source* Studies of the Tokaido Shinkansen, vol. 1, RTRI, 1960, p. 113

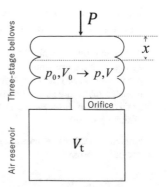

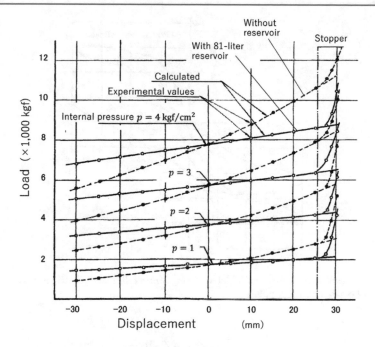

Fig. 4.39 Static characteristics of the air spring used in the "Asakaze." *Source* Studies of the Tokaido Shinkansen, vol. 1, RTRI, 1960, p. 115

4.5.1.2 Lateral Characteristics of Air Spring

By utilizing the air spring's lateral elasticity, an attempt was made to abolish swing hanger systems, which had complicated structure and were prone to friction and play. In 1959, the first prototype bogie of this idea was built for conventional electric railcars, and in 1961, it was decided that this system would be used for the Shinkansen bogie.

However, as shown in Fig. 4.40, the lateral characteristics of air springs at that time included nonlinearity and hysteresis. That is, spring extension/contraction was not proportional to the applied force, and when a car body was displaced by lateral force, it did not entirely return to its original position even after the force was removed.

Since the shear deformation of the bellows caused the nonlinearity and hysteresis, efforts were made to overcome these defects. The resulting characteristics, as shown in Fig. 4.41, still had somewhat more hysteresis but much-improved linearity, so this type of air spring was introduced in the Shinkansen test cars for the test section.

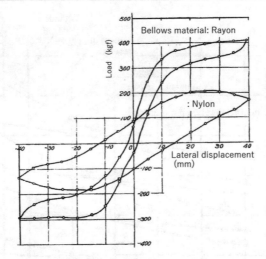

Fig. 4.40 Lateral characteristics of initial air spring. *Source* Research on High-Speed Railways, Kenyusha Foundation, 1967, p. 301

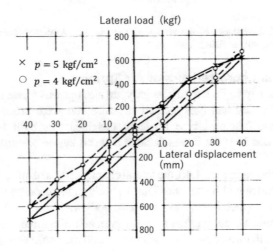

Fig. 4.41 Lateral characteristics of improved air spring. *Source* Research on High-Speed Railways, Kenyusha Foundation, 1967, p. 303

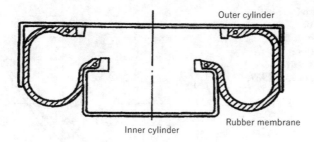

Fig. 4.42 Cross section of the diaphragm-type air spring *Source* Studies of the Tokaido Shinkansen, vol. 5, RTRI, 1964, p. 347

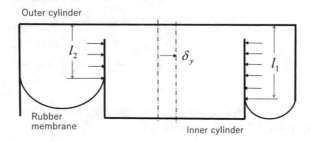

Fig. 4.43 Principle of lateral spring action. *Source* Studies of the Tokaido Shinkansen, vol. 5, RTRI, 1964, p. 347

In this way, the bellows-type air spring was improved, and its characteristics were considerably better, but the hysteresis problem remained. This shortcoming was fundamentally remedied in 1962 by a new type of diaphragm air spring with a cross section shown in Fig. 4.42 devised by Sumitomo Metal Industries and Sumitomo Electric Industries. Hysteresis was almost eliminated, and this type of air spring was used in the Shinkansen mass production car.

The lower part is connected to the auxiliary tank through an orifice (not shown in the figure), and the principle of vertical spring action is the same as the bellows type.

The principle of lateral spring action in the diaphragm type is entirely different from that of the bellows type. As shown in Fig. 4.43, when the inner cylinder moves to the right, the left and right rubber films deform. In this case, the force proportional to l_1 works from right to left, and the force proportional to l_2 works from left to right, so the resultant force proportional to $(l_1 - l_2)$ acts as the spring force.

Fig. 4.44 Lateral
characteristics of
diaphragm-type air spring.
Source Research on
High-Speed Railways,
Kenyusha Foundation, 1967,
p. 304

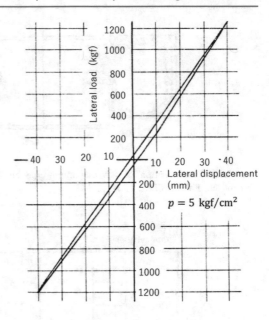

As shown in Fig. 4.44, the diaphragm-type air spring had superior characteristics in linearity and hysteresis to the bellows type.

4.5.2 Ride Quality of Pre-Mass-Produced Cars

Figures 4.45 and 4.46 show the ride quality of the six pre-mass production cars,[10] which composed Train C, on the test section in 1963. As shown in Fig. 4.45, most of the cars' ride quality was under 1.5 at 210 km/h, a passing score for each vehicle. However, as Fig. 4.46 shows, the ride quality for some cars ranged from 2 to 3 for lateral vibration and 1.5 to 2 for vertical vibration, indicating that further improvement was required to achieve higher speeds.

---

[10] Pre-mass production cars are the vehicles that allow railway companies to find problems before they mass-produce. Pre-mass production cars come after prototypes or test cars.

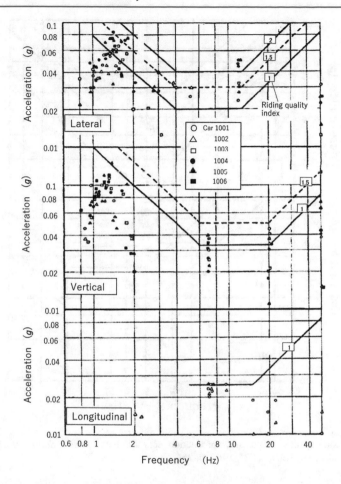

Fig. 4.45 Ride quality of pre-mass production cars (200–210 km/h). *Source* Studies of the Tokaido Shinkansen, vol.5, RTRI, 1964, p. 356

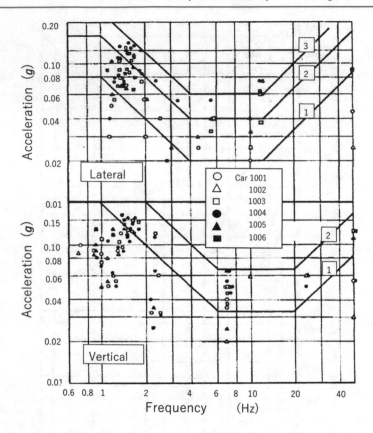

Fig. 4.46 Ride quality of pre-mass production cars (230–246 km/h). *Source* Studies of the Tokaido Shinkansen, vol.5, RTRI, 1964, p. 356

4.6 Brakes

Stopping a train of mass M and speed V requires absorbing or consuming the train's kinetic energy $MV^2/2$. The recipients of this kinetic energy are usually adhesive brakes and train's running resistance. The adhesive brake is only applicable in the range where the wheels do not slide, so its ability depends on the adhesion coefficient.[11]

When the Shinkansen project started, RTRI had no data on adhesion coefficients at speeds of 160 km/h or higher, so initially, in addition to adhesive brakes, it considered the use of nonadhesive brakes such as electromagnetic rail brakes and aerodynamic brakes.

---

[11] Adhesive coefficient is the value obtained by dividing the adhesive force with which the wheel begins to slide by the wheel load at rest.

However, electromagnetic brakes were not adopted because basic tests showed that they required a large iron core that might affect the conventional bogie structure in trying to obtain sufficient braking force. Meanwhile, the Shinkansen, which has no level crossing with roads and good visibility, does not require the same short braking distance (the distance from where the train applied the brakes to where it stopped) as conventional railways, so only adhesive brake systems were adopted.

For adhesive brake systems, it was decided to use electric brakes and friction brakes, and for friction brakes, a disk brake system was adopted. Because wheel treads are essential for the Shinkansen vehicle's running stability, roughening the surface or raising the temperature of treads by brake shoes had to be avoided.

4.6.1   Adhesion Limitation and Adhesion Coefficient

As braking torque is applied to wheels, adhesion strength increases correspondingly, but sliding occurs once it reaches a specific value. In 1959, to get an idea of this limitation of adhesion around 200 km/h and predict braking distance at high speeds, experiments were conducted using test apparatus. The apparatus consisted of two 400-mm-diameter cylinders with a contact width of 20 mm; one cylinder served as track and the other as wheel. The apparatus could change speeds up to 250 km/h, wheel loads up to 20 kgf/mm, and surface conditions to wet or dry.

Figure 4.47 shows the experimental results with the speed at which gliding started on the horizontal axis and the adhesive coefficient on the vertical axis.

The results were as follows:

- Adhesion coefficient decreased continuously as speed increased.
- Coefficients of wet surfaces were about half of those of dry surfaces.
- The coefficient at 220 km/h was about 72% of that at 100 km/h.

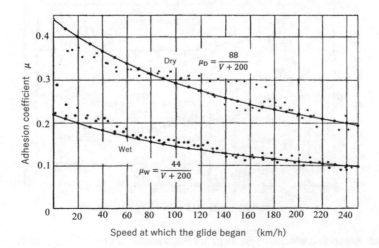

Fig. 4.47 Adhesion coefficients of the test apparatus. *Source* Studies of the Tokaido Shinkansen, vol. 1, RTRI, 1960, p. 131

The experiments yielded the following empirical formulas for the adhesive coefficients μ_D and μ_W for dry and wet surfaces:

$$\mu_D = \frac{88}{V+200}, \mu_W = \frac{44}{V+200} \quad (V : \text{km/h})$$

However, these results could not be applied to actual brake systems because the contact surface between the wheel and the rail was 1/4 of the real size. Therefore, in 1960, a large adhesive test machine with a contact ratio of 1:1 was installed. Newly measured adhesion coefficients obtained from the large machine yielded the following empirical formula, which can be applied to actual cars:

$$\mu_W = \frac{44.7}{V+147} - 0.027$$

Curves ① and ② in Fig. 4.48 are the same lines as those shown in Fig. 4.47, and ③ is the curve obtained using the new test machine.

By using the large testing machine, the adhesion coefficient at high speed was obtained, but since various adverse conditions would occur in actual vehicles, the results obtained by the experimental device could not be used as they were. Therefore, for real cars, the adhesion coefficients of dry and wet surfaces, μ_D and μ_W, were finally determined as follows and allowed for the uncertainty of long-term use.

$$\mu_D = \frac{27.2}{V+85}, \mu_W = \frac{13.6}{V+85}$$

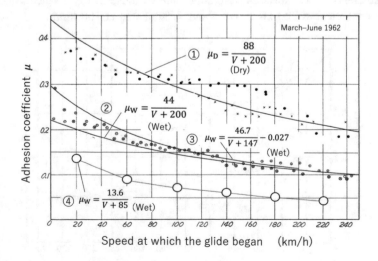

Fig. 4.48 Adhesion coefficients with large testing machine. *Source* Research on High-Speed Railways, Kenyusha Foundation, 1967, p. 338

In Fig. 4.48, ④ shows the line calculated by $\mu_W = \frac{13.6}{V+85}$ in wet conditions. The brakes of the Shinkansen train were designed based on the assumption that adhesion coefficients would be greater than ④ in Fig. 4.48 even under a variety of adverse conditions.

4.6.2 Disk Brakes

If the electric brakes fail in an emergency, the disk brakes alone must stop the train. In this case, the amount of energy that the disk brake system must absorb is enormous and requires many disks. As a result of examining the place to install the disks, it was decided that they should be placed on both sides of wheels.

The selection of disk and pad material was carried out using a brake testing machine that can simulate conditions similar to those of actual trains (Photo 4.20).

Two types of cast steel and four types of cast iron with different chemical compositions were tested for the disk in terms of compatibility with mating pads, strength under high temperature, and wear resistance, and cast iron was selected as the disk material. Nine types of pads, including iron-based sintered alloy, copper-based sintered alloy, resin-based synthetic material, and ceramic-based pads were examined, and copper-based sintered alloy, which is mainly composed of 60–70% copper and 5–15% tin, was selected. Iron-based pads were eliminated because they could damage the iron-based disk.

Figure 4.49 shows that the disk temperature would rise to 300 °C when a train stops from 200 km/h with the disk brake system only.

Photo 4.20 Brake testing equipment. (Provided by RTRI)

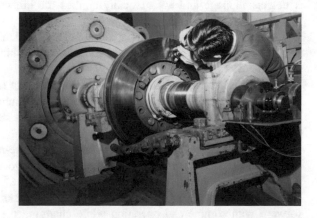

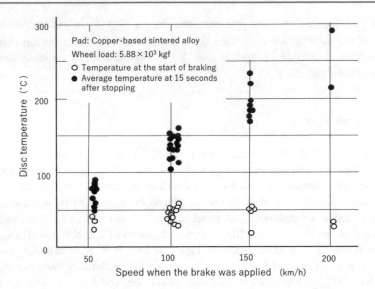

Fig. 4.49 Increase in disk temperature. *Source* Research on High-Speed Railways, Kenyusha Foundation, 1967, p. 326

4.6.3 Braking Tests on the Test Section

In braking tests on the test section, wheel locks often occurred due to rust and dust on rail surfaces, so tests continued by reinforcing anti-skid devices and equipping all wheels with tread cleaning devices. The anti-skid device releases the brake when it detects slippage and automatically applies the brakes again after adhesion has recovered.

Figure 4.50 shows the disk temperature of six-car Train C as measured in March 1964; the maximum temperature was about 300 °C, which is almost the same as the experimental value shown in Fig. 4.49.

Figure 4.51 shows the adhesion coefficients of wet rails measured five months after the opening. Considerable variation in the measurements suggests that contact surfaces were not yet stable. The line marked "Design standard (wet)" in the figure is the same as ④ in Fig. 4.48.

Figure 4.52 shows the braking distance of the pre-mass-produced Train C. It indicates that trains traveling at a speed of 200 km/h would need 2000 m to come to a stop even if it applied the emergency brake.

For the deformation and cracks that occur in disks, it was decided to move forward with commercial operation while changing the usage limits for the time being.

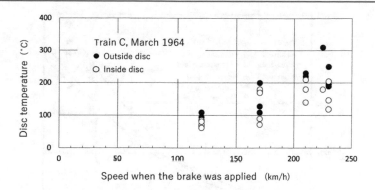

Fig. 4.50 Disk temperature. *Source* Research on High-Speed Railways, Kenyusha Foundation, 1967, p. 326

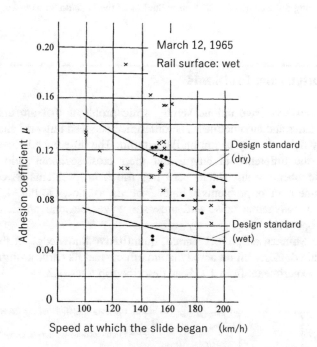

Fig. 4.51 Adhesion coefficient after the Shinkansen opening. *Source* Studies of the Tokaido Shinkansen, vol. 6, RTRI, 1965, p. 21

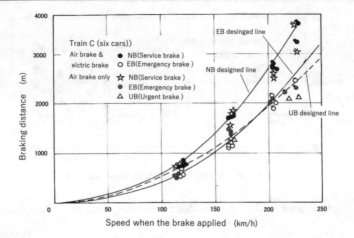

Fig. 4.52 Braking distance of Train C. *Source* Studies of the Tokaido Shinkansen, vol. 5, RTRI, 1964, p. 25

4.7   Aerodynamic Problems

High-speed railways have unique aerodynamic problems that are different from those of airplanes and automobiles. The difference between railways and aircrafts is whether they fly in the sky or run on the ground. The difference between railways and cars is the difference in length and their cross-sectional ratio to tunnels. Aerodynamic phenomena, which were not much of problems at conventional speeds, became more of problems at higher speeds, as shown in Fig. 4.53. Of these phenomena, aerodynamic noise and pressure waves became problems after the Shinkansen opening.

When the Shinkansen project started, quantitative knowledge of these matters was very limited. Research on aerodynamic effects had its earliest origin in Miki's wind-tunnel experiments from 1954, as described in Sect. 1.2.

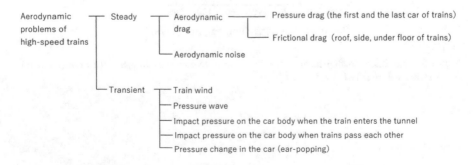

Fig. 4.53 Aerodynamic problems in high-speed railways

The following sections look at some of the aerodynamic problems that were revealed in the process of realizing the Shinkansen.

4.7.1 Body Shape

When a train is running, its front is subjected to wind pressure (pressure resistance at the front), the rear end is pulled backward (eddy resistance at the rear), and the sides, bottom, and roof surfaces are dragged by the surrounding air. The aerodynamic drag D of the vehicle made up of these factors is expressed as:

$$D = \frac{1}{2}\rho v^2 C_x A$$

where ρ is air density, v is wind speed, A is the cross-sectional area of the vehicle, and C_x is the drag coefficient (the sum of the drag coefficients created by the aforementioned factors).

To reduce the drag, it is necessary to make the head streamlined, reduce the cross-sectional area of the car, smooth the seams of the train with outer hoods, add bogie covers, and make the additives (such as vents and pantographs) streamlined and as small as possible.

Wind-tunnel tests on train aerodynamic drag were first carried out as far back as 1937 when the RTRI was still the Railway Ministry's Office Research Institute.

As shown in Fig. 4.54, drag coefficients of scale model cars were measured for four types (scale 1/25), including the shapes of front heads, presence/absence of air vents on roofs, and presence/absence of skirts, with the results that when the drag coefficient of No. 1 was 100%, then No. 2 was 70%, No. 3 was 63%, and No. 4 was 45%.

Miki's plan of a four-and-a-half-hour Tokyo–Osaka trip, which appeared in a newspaper in October 1953 and led to Miki receiving a research grant, seems to have referred to this test data.

His wind-tunnel testing, funded by the grant, began in earnest in 1954. He built seven different types of 1/10 scale model railcars and one seven-car train in 1/40 scale. The latter was intended to study the relationship between air drag and train length. Wind-tunnel experiments were carried out eight times from 1954 to 1961. Miki left behind many papers on the subject. Some of the results are shown in Fig. 4.55.

The features of the models used in the experiments are as follows:

Type SH: A scale model of the Series 80 train (Photo 1.4), introduced in 1950.

Type SH*: SH with a head used in German diesel cars.

Type SE: A model of a high-speed car with a streamlined head, low roof, rounded body, and bogie cover.

Type SE*: A model with a longer head than that of the SE.

Type SE-SH: SE with the head of the SH.

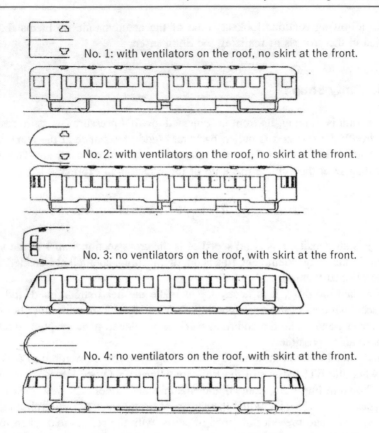

No. 1: with ventilators on the roof, no skirt at the front.

No. 2: with ventilators on the roof, no skirt at the front.

No. 3: no ventilators on the roof, with skirt at the front.

No. 4: no ventilators on the roof, with skirt at the front.

Fig. 4.54 Models of first wind-tunnel test. *Source* "On the Results of Wind Tunnel Tests of Stream-Lined Rolling Stock Models", Research Materials, no. 2, Railway Minister's Office Research Institute, 1937, p. 33

Type SK: The type with a raised cab floor.

Type F: SE-SH with a flat face.

Shinkansen test car: The scale model of the test car for the Shinkansen test section.

Odakyu's Type 3000 SE train, introduced in 1957, took its base from Type SE. The Series 0 Shinkansen vehicle was a modified version of Type SK (Photos 4.21 and 4.22).

In his paper, Miki stated [14]:

In the case of Type SH, the drag coefficient was reduced to about 70% when the frontal area of the head was as streamlined as is generally practiced in Europe. In the case of Type SE, the coefficient decreased to 35% of that of SH. This is a remarkable phenomenon. (Author's translation of the Japanese).

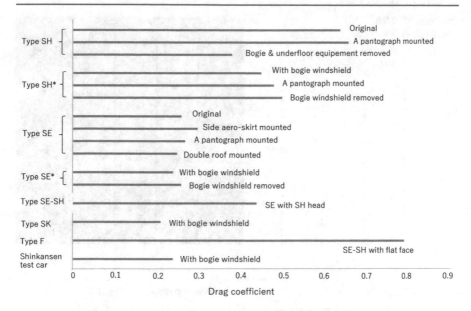

Fig. 4.55 Drag coefficient of the models

Photo 4.21 1/10 scale models, from left to right: Type SE*, SK, Kodama, and F. (Provided by Shinichi Tanaka)

He listed future issues, including an increase in drag due to longer trains and an increase in impact wind pressure when entering tunnels at high speed or passing oncoming trains.

The breakdown of the drag coefficients for Types SH and SE shown in Fig. 4.56 indicates that the aerodynamic drag of conventional railcars can be significantly reduced by improving the cars' shape. In the figure, the frictional drag of Type SE accounts for only about half of the total because it is only for the lead car. However, as frictional drag increases with train length, it is essential to reduce dragged air in longer trains. That is, it is necessary to make boundary layers thinner.

Photo 4.22 1/10
Wind-tunnel test[12]

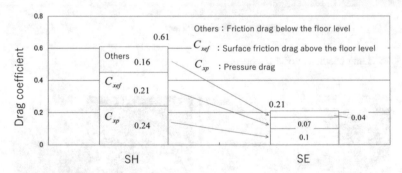

Fig. 4.56 Breakdown of drag coefficient for Types SH and SE

Figure 4.57 shows the growth of boundary layers when trains become longer. It indicates that when the distance from the train head is x, the boundary layer thickness δ is expressed by the following empirical equation. The marks □ in the figure are the wind-tunnel data using the Type SE 1/40 scale model. The marks ◆ are the values obtained by the 3000 series SE train after its business opening.

$$\delta = 0.24x^{2/5}$$

Based on the aforementioned experiments, the Shinkansen test car's shape was determined by modifying the SK model that had the lowest aerodynamic drag in Fig. 4.55, and its wind-tunnel test was conducted at the end of 1961 for a 1/12 scale model (Photo 4.23).

---

[12] Miki, T. et al., "Research on Aerodynamics of High-Speed Railcars," *Journal of the Japan Society of Mechanical Engineers* (JSME), no. 478, 1958, p. 36.

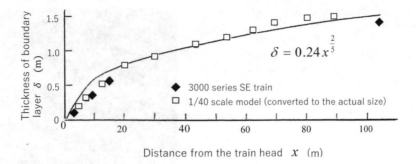

Fig. 4.57 Boundary layer growth. *Source* Miki, T. et al., "Research on Aerodynamics of High-Speed Railcars", Journal of the Japan Society of Mechanical Engineers, no.478, 1959, p. 40.

Photo 4.23 Wind-tunnel test of 1/12 scale Shinkansen lead car model. (Provided by RTRI)

As shown in Fig. 4.55, the Shinkansen lead car model's drag coefficient was 0.24 (0.25 for the SE and 0.21 for the SK). This value was a little smaller than that of the SE, but the Shinkansen car's cross section was 1.5 times that of the SE cars, and Shinkansen trains were far longer than SE trains, so the resultant drag force of Shinkansen trains was expected to be 50 to 60% larger than that of the SE.

Photo 4.24 shows a wind pressure brake test model.

Photo 4.24 Model for wind pressure brake[13]

4.7.2   Transitional Aerodynamic Problems

When the Shinkansen project started in 1958, RTRI had little knowledge of the transient aerodynamic problems that occur when trains enter tunnels or pass each other. The first measurement was made in September 1958 when a conventional train entered a tunnel. Instruments that could keep up with the fast air pressure fluctuation were not available, so the Shinkansen team made its own measuring instruments.

4.7.2.1   Aerodynamic Phenomenon When a Train Enters a Tunnel

Figure 4.58 shows the pressure changes at the train head and inside the car as a train enters and exits a 1.8-km single-track tunnel at approximately 80 km/h. As shown in the figure, when the train enters the tunnel, the pressure at the train face jumps to 3.5–4 times its original level and then drops rapidly. It then continues to fluctuate and decrease, and as the train exits the tunnel, it rises slightly and then returns to its original value. In a double-track tunnel, the impact wind pressure was only twice as high as outside the tunnel, and the subsequent variation was monotonous compared to the single-track tunnel.

---

[13] Miki, T., "Problems Related to High-Speed Railway," *Journal of the Japan Society of Mechanical Engineers* (JSME), No. 480, 1959, p. 143.

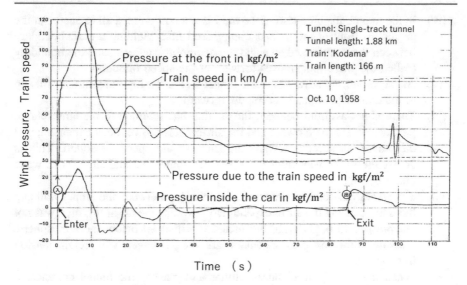

Time (s)

Fig. 4.58 Pressure change in a tunnel. *Source* Ito, H., et al., "Wind Pressure Measurement of High-Speed Trains", RTRI Preliminary Report, no.36, RTRI, 1959, p. 10.

(1) **Mechanism of Pressure Fluctuation**

In describing the pressure fluctuation phenomenon, Tomoshige Hara,[14] an expert in fluid dynamics, explained the pressure fluctuation that a train experiences while passing through a tunnel as follows [15]:

(i) There are two causes of air pressure fluctuation when a train enters a tunnel: one is the fluctuation caused by the compression of the air in the tunnel, and the other is the one caused by the viscosity of the air when the air in the tunnel is blown out of the entrance through the gap between the tunnel wall and the train.

(ii) When the train enters the tunnel, the air that the train excludes has to escape somewhere. Most of it blows outward from the tunnel entrance in reverse. The blown-out air volume depends on the train speed and the cross-sectional area ratio of the tunnel to the train. For the Shinkansen train at 200 km/h, the air volume blown out will be about 80% of the train volume.

(iii) Air not blown out at the entrance is compressed in front of the train. Therefore, if the train speed is high or the tunnel cross-sectional area is small, the percentage of air compressed in front of the train increases, and the pressure rise in front of the train is large.

---

[14] Tomoshige Hara joined the Imperial Japanese Navy in 1936, moved to RTRI in 1945 as a senior researcher, and later became head of Hara Special Laboratory of RTRI.

(iv) In the tunnel, the air ahead of the train is not compressed all at once, but the air just in front of the train is compressed first, yielding a clear boundary between the compressed area and the uncompressed area, and the boundary (called the wavefront) moves forward at the speed of sound.

(v)   As the wavefront advances to the tunnel's exit, the high-pressure air fills up to the exit of the tunnel and blows forward.

(vi) There is a wavefront between the air that has returned to the atmospheric pressure due to the blowout and the air that has not yet been expanded, and the wavefront recedes toward the train at the speed of sound. This is equivalent to the wavefront reflects at the tunnel exit and changes its wave phase by 180°.

(vii) The wavefront continues to recede into the gap between the tunnel wall and the train, causing the air to expand. However, as the air in this area was not compressed to begin with, the pressure will drop below the atmospheric pressure. If this low pressure enters cars, the passengers will feel the tension in their ears.

(viii) When the wavefront advances further and reaches the tunnel entrance, it reflects there again, restoring the negative pressure to the atmospheric pressure as it returns.

Based on his understanding, Hara explained the general pattern of pressure change in the train's frontal area in the tunnel, as shown in Fig. 4.59.

While a train is traveling at a speed of U outside tunnels, the train's front pressure is $\frac{\rho_0 U^2}{2}$, as shown in the figure. However, as the train head enters the tunnel, the pressure jumps up to the mark G by the compressed air in front of the train. The pressure increase from G to H is caused by the train pushing the air outward the tunnel entrance. Once the train's rear end enters the tunnel, this pressure rise disappears because there is no longer a need to push air out of the tunnel. The wave-like fluctuations after the mark H in the figure are due to the wavefront moving through the tunnel with repeated reflections. This pattern takes various forms depending on the length of the tunnel, cross-sectional ratio of the train to the tunnel, train size, train speed, and tunnel wall's condition.

Figure 4.60 shows the pressure changes in the leading car when the four-car Train B (100 m long) entered a 471-m double-track tunnel at 200 km/h, and when a five-car conventional train (125 m long) entered a 1055-m single-track narrow gauge tunnel at 79 km/h. In both cases, the mark A shows when the train's front entered the tunnel, B is when the train's rear end entered, and C is when the wavefront reflected at the tunnel exit reached the train's front. Part (a) of the figure shows that the wave-like pressure fluctuations after C are due to the reflected waves that come many times. Although the two patterns appear to be unrelated at first glance, they are essentially the same phenomenon.

Figure 4.61 shows the pressure fluctuation in a tunnel on the Shinkansen test section when Train B entered the aforementioned tunnel at 198 km/h. The wide oscillogram in part (b) of the figure, shown in black, is due to irregular pressure

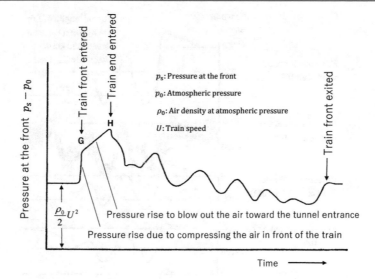

Fig. 4.59 General pattern of pressure variation in the frontal area in a train. *Source* Studies of the Tokaido Shinkansen, vol. 1, RTRI, 1960, p. 92.

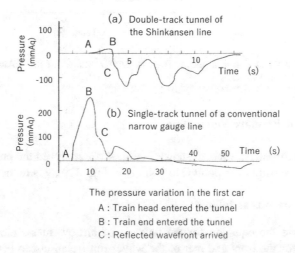

The pressure variation in the first car

A : Train head entered the tunnel
B : Train end entered the tunnel
C : Reflected wavefront arrived

Fig. 4.60 Two patterns of atmospheric pressure variation. *Source* Research on High-Speed Railways, Kenyusha Foundation, 1967, p. 361

oscillation caused by airflow turbulence at the train's rear end. Part (a) in the figure indicates that the pressure rise due to viscosity is relatively small because of the short length of the train (corresponding to G to H in Fig. 4.59), and the variation due to the reflection of the pressure wave is greater.

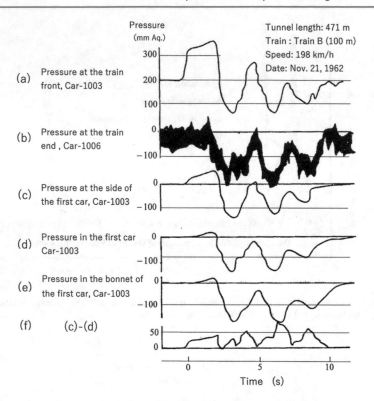

Fig.4.61 Pressure fluctuation in a tunnel, Train B. *Source* Studies of the Tokaido Shinkansen, vol. 4, RTRI, 1963, p. 135

(2) **Analysis of Pressure Fluctuation**

Hara, using the analysis model shown in Fig. 4.62, analyzed the pressure change due to the air viscosity from point G to point H in Fig. 4.59, where the pressure is at its maximum.

The procedure was as follows:

(i)   Formulate the equation for the balance of airflow mass, momentum, and energy for the front and rear of the wavefront (relationship between ⓪ and ①).

(ii) Formulate the equation for the balance of airflow mass, momentum, and energy at the front of the train (the relationship between ① and ②).

(iii) Formulate the equation for the change in momentum to the air column between the front of the train and the tunnel entrance (the relationship between ② and ③).

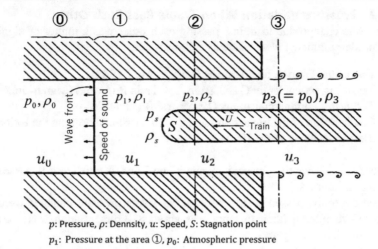

p: Pressure, ρ: Dennsity, u: Speed, S: Stagnation point

p_1: Pressure at the area ①, p_0: Atmospheric pressure

Fig. 4.62 Pressure variation analysis model. *Source* Research on High-Speed Railways, Kenyusha Foundation, 1967, p. 360

Fig. 4.63 Comparison of measured and calculated values. *Source* Research on High-Speed Railways, Kenyusha Foundation, 1967, p. 360, 364

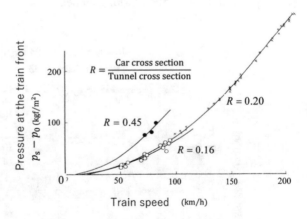

He solved these equations to obtain the relationship between the train speed and the pressure change at the train front. The calculated value agrees well with the measured value as shown in Fig. 4.63 (vertical axis: $P_s - P_0$ [the difference between the pressure at the train front and atmospheric pressure], horizontal axis: train speed).

4.7.2.2 Pressure Variation When Trains Pass Each Other

The pressure change due to trains' passing each other was a matter of significant concern. Measurements were made as follows:

(i)   Test using Train A and Train B (March–April 1963)
(ii)   Test using six-car Train C and Train A + Train B (6 cars) (March–June 1964)
(iii)   Test using commercial trains (12-car train) (July 1964)
(iv)   Confirmation of the impact pressure due to the Shinkansen on the Keihanshin Electric Railway (July 1964).

Figure 4.64 shows the pressure change when Train A and Train B passed each other outside tunnels.

When the trains passed each other, the car side's pressure changes were + 85 mmAq to –70 mmAq (their distance was about 820 mm), which was not a particularly problematic figure.

However, the pressure changes when they passed each other in a tunnel were quite different from those outside the tunnel due to the pressure wave reflection, as described previously.

Figure 4.65 shows the pressure variation measured by 12-car commercial trains passing each other in a 5008-m-long tunnel.

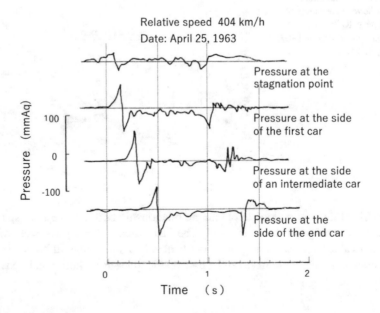

Fig. 4.64 Pressure changes when trains pass each other outside tunnels. *Source* Research on High-Speed Railways, Kenyusha Foundation, 1967, p. 385

Fig. 4.65 Pressure change when trains pass each other in a long tunnel. *Source* Research on High-Speed Railways, Kenyusha Foundation, 1967, p. 385

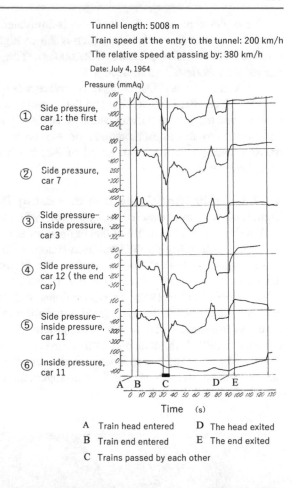

Tunnel length: 5008 m

Train speed at the entry to the tunnel: 200 km/h

The relative speed at passing by: 380 km/h

Date: July 4, 1964

Pressure (mmAq)

① Side pressure, car 1: the first car

② Side pressure, car 7

③ Side pressure– inside pressure, car 3

④ Side pressure, car 12 (the end car)

⑤ Side pressure– inside pressure, car 11

⑥ Inside pressure, car 11

Time   (s)

A   Train head entered   D   The head exited
B   Train end entered   E   The end exited
C   Trains passed by each other

The large negative pressure at the time of passing each other (point C in the figure) was the result of a combination of the pressure trough caused by the pressure wave reflection due to a train's entry into the tunnel and the pressure trough due to the trains passing each other. These values, which varied in waveforms and peak values depending on the relative positions of both trains in the tunnel, were summarized as follows:

(i)   The maximum value of car side pressure is about + 202 mmAq to –471 mmAq.

(ii)   The maximum value in an airtight car is about + 25 mmAq to –110 mmAq.

(iii) The pressure fluctuation in an airtight car is gradual, so passengers will feel only slightly uncomfortable.

(iv)   There is no problem with the cab's front glass's maximum stress value, which is approximately 0.75 kgf/mm^2 in both positive and negative pressures.

The negative pressure of 470 mmAq is equivalent to 470 kgf per square meter, which means that the car's side, which is 2.3 m high and 25 m long, is subject to instantaneous negative pressure of 27,000kgf. The car frames were strengthened against this, as described in Sect. 4.4.

Between Kyoto and Osaka, there is a place where the Shinkansen and a private railway are parallel to each other and their windows are only 3 m apart. There was concern that the private railway would be adversely affected when the trains passed each other, so the impact pressure on the private railway was measured. The pressure changes at a relative speed of 322 km/h were within ± 20 mmAq, an acceptable level.

4.7.2.3 Train Wind (Wind Due to Passing Trains)

Train wind was first measured in September 1957 for SE trains. The result was that the wind speed was about 5% of that of the train at a distance of about 1 m from the car side at 100 m behind the train head (Photo 4.25).

In response to the decision to construct the Shinkansen, it became necessary to determine the distance between the two tracks and confirm the ground workers' safety; in November 1958, 25 conventional railcars were connected to form a 500-m train, for which the wind speed distribution in the boundary layer around the train was measured on-board and on the ground. On-board measurement was made on the 1st, 10th, 11th, 20th, and 24th car using a 2-m-long measuring bar. The bar had 15 pitot tubes and was mounted perpendicular to each car's side. Measurements on the ground were done using an arrangement of anemometers, as shown in Fig. 4.66.

Photo 4.25 Train wind measurement by Odakyu SE train. (Provided by RTRI)

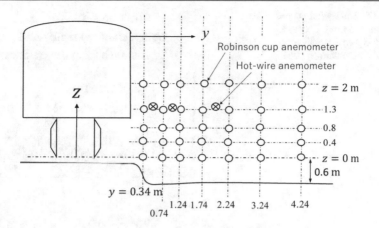

Fig. 4.66 Arrangement of anemometers. *Source* "Train Wind Measurement of 25-Car Train", RTRI Preliminary Report, no.91, RTRI, 1959, p. 7

The measurements showed that the train wind was zero at 70 cm from the first car's side, about 20% of the train speed at 2 m from the 24th car's side, and about 10% at 4 m from the train side 450 m behind the train head.

The next measurements were made in July 1959 on a six-car train at 160 km/h; the results were almost identical to those mentioned in the preceding paragraph. However, as the train speed increased, the anemometer's response became a problem. With its rotational momentum and friction, the Robinson cup anemometer was suitable for measuring the average wind speed, but it was not suitable for instantaneous wind changes. The hot-wire anemometer was ideal for measuring mild winds, but not for strong winds such as train winds.

By the time the Shinkansen test runs began, the aforementioned problems with anemometers seemed to have been solved, and the hot-wire anemometers, with better response, were used on the test section.

Figure 4.67 shows the train wind measured at a height of 1.6 m above the rail level at the test section's embankment, with the time on the horizontal axis. The point I is when the train head passed; II is when the rear end passed. The vertical axis is the ratio of train speed U_0 to wind speed U. Since the measurements varied, a large number of measurements were superimposed to read the average, as shown in the figure. The figure shows that the maximum train wind occurs after the rear end has passed. In the figure, the maximum U/U_0 is about 0.3 at a distance of 2.96 m from the track center (1.27 m from the car's side), indicating that, at this location, the train wind is about 18 m/s at a train speed of 210 km/h (58 m/s).

Fig. 4.67 Train wind caused by Train B. *Source* Research on High-Speed Railways, Kenyusha Foundation, 1967, p. 388

U_0: Train speed

U: Speed of the train wind

y: Distance from the track center

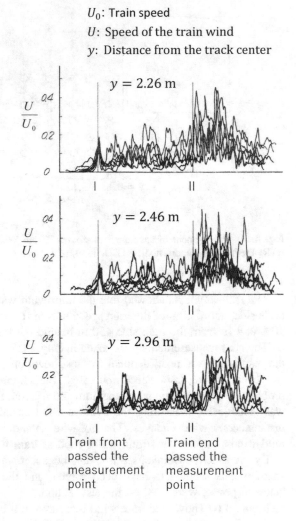

I Train front passed the measurement point

II Train end passed the measurement point

4.7.2.4 Countermeasures Against Pressure Change in Cars

A new problem that arose after test runs began was the so-called ear-popping, a discomforting phenomenon. The ear's tympanic membrane feels the pressure difference between the two sides, and as the pressure difference increases, the discomfort becomes painful. However, the middle ear behind the eardrum is connected to the outside air through the Eustachian tube. When a large pressure difference occurs between the two sides of the eardrum, the pressure difference is resolved through this Eustachian tube. Therefore, if the pressure change's absolute value is large but the pressure change is not rapid, the Eustachian tube's pressure regulation function will reduce or eliminate discomfort.

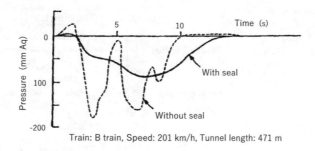

Fig. 4.68 Air pressure change in a sealed car. *Source* Studies of the Tokaido Shinkansen, vol. 4, RTRI, 1963, p. 14

In January 1963, a test was conducted on Train B which involved removing air-conditioning units of one car, blocking the gaps leading to the outside with adhesive tape, and sealing the sliding doors of doorways.

Figure 4.68 shows the pressure change inside the car at the test. Since the sealing was not perfect, there was some air leakage, and the absolute value of the pressure change was only about half of what it was before the sealing. However, the speed of the pressure fluctuation became so small that the ear discomfort was almost imperceptible. Akiya Yamamoto[15] examined the relationship between the gap leading to the outside and the pressure change inside the vehicle and found the following relationship:

$$\sqrt{|p_0 - p_1|} - \sqrt{|p - p_1|} = 2St/V$$

where p_0 is the pressure outside the car in mmAq, p_1 is the inside pressure at the time t in mm Aq, S is the area of the openings in cm^2, V is the volume of the car in m^3, and t is time in seconds.

From this equation, Yamamoto deduced that the gap left in the car in Fig. 4.68 would have been 85cm^2, and he said that openings of this size would be acceptable in Shinkansen cars.

Figure 4.69 shows the air pressure change in the airtight six-car Train C composed of pre-mass production cars as measured in March 1964.

The outside pressure fluctuated significantly, as shown in the figure, but since the car was now airtight, the inside pressure fluctuation was very low and ear discomfort disappeared.

The dotted line in the second figure from the top shows Train B's pressure fluctuation; Train B was not airtight.

---

[15] Akiya Yamamoto joined JNR in 1955. He later became head of the physics laboratory of RTRI after served as a senior researcher.

Fig. 4.69 Change in air pressure in Train C. *Source* Studies of the Tokaido Shinkansen, vol. 5, RTRI, 1964, p. 237

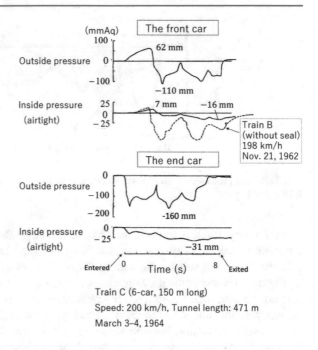

Train C (6-car, 150 m long)

Speed: 200 km/h, Tunnel length: 471 m

March 3–4, 1964

4.8 Running Resistance

Running resistance of trains is an essential quantity for the design of traction and braking systems. It consists of mechanical drag D_m and aerodynamic drag D_a. The former is due to friction and deformation of mechanical components, and the latter is due to aerodynamic factors, as described in Sect. 4.7.

In general, the running resistance R is expressed as follows.

$$R = D_m + D_a = (a + bV)W + CV^2 \tag{4.1}$$

where R is listed in kgf, V is train speed in km/h, W is train weight in tons, C is aerodynamic drag coefficient, and a, b are constants determined by train running tests.

4.8.1 Assumption of Running Resistance at High Speeds

When the Shinkansen project started, there was no data on train's running resistance above 130 km/h in Japan. Therefore, in the design of the Shinkansen test vehicles, Train A and Train B, the following empirical equation was used to assume the running resistance based on data up to 130 km/h:

$$R = (1.6 + 0.035V)W + CV^2 \tag{4.2}$$

For the four-car Train B, a cross-sectional area of 8.88 m², a length of 100 m, and weight of 240 tons were applied to obtain the following equation:

$$R = (1.6 + 0.035V)240 + 0.02933V^2 \tag{4.3}$$

4.8.2 Measurement of Running Resistance

Measurement of the running resistance of the test trains began in November 1962. Figure 4.70 shows the results for the four-car Train B. Since the measurements were almost the same as those calculated by Eq. 4.3, Eq. 4.2 was applied to the 12-car Shinkansen train.

The formula for the 12-car train was as follows:

$$R = (1.6 + 0.035V)W + 0.0486V^2 \tag{4.4}$$

In July 1964, two months before the start of business operations, 12-car train's running resistance was measured and the results were as shown in Fig. 4.71. Since the measurements were a little below the line calculated by Eq. 4.4, the formula was slightly modified to the formulas shown in Eqs. 4.5 and 4.6. The dotted line in Fig. 4.71 is based on Eq. 4.4, the two solid lines are based on Eqs. 4.5 and 4.6, and the dashed line is the trend line of running resistance calculated from the data on power running.

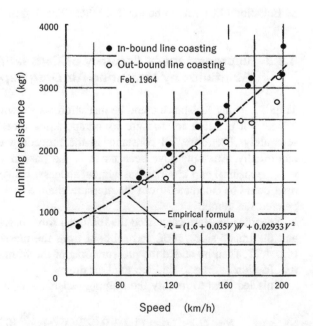

Fig. 4.70 Running resistance of the four-car Train B. *Source* Research on High-Speed Railways, Kenyusha Foundation, 1967, p. 166

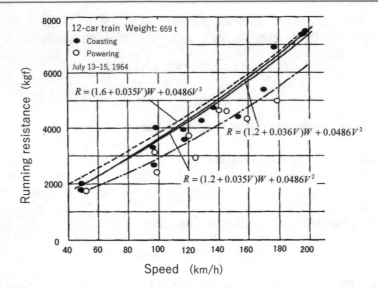

Fig. 4.71 Running resistance of 12-car train. *Source* Research on High-Speed Railways, Kenyusha Foundation, 1967, p. 166

$$R = (1.2 + 0.035V)W + CV^2 \qquad (4.5)$$

$$R = (1.2 + 0.036V)W + CV^2 \qquad (4.6)$$

Equation (4.6) was to be applied when focusing on higher speeds.

4.8.3 Improvement in Accuracy of Calculating Running Resistance by Measuring Air Drag Separately

Train's running resistance could be measured, as shown in Figs. 4.70 and 4.71, but it was not possible to measure its components, mechanical drag, and air drag, separately. Therefore, all constants in the preceding equations were determined empirically. Attempts have been made in the past to separate the total resistance into mechanical and air drag, but without success. However, Hara found that the air drag could be calculated from the pressure change at the train front when the train entered the tunnel.

Specifically, Hara calculated the frictional force between the air around the train and the train's sides, roof, and subfloor from the pressure change from G to H in Fig. 4.72. Then he added the pressure drag at the front and rear end of the train to this frictional force to find the total air drag.

This led Hara to modify the running resistance of a train as follows:

$$R = D_m + D_a = (1.2 + 0.022V)W + \frac{\rho A}{2}(0.2 + 0.0045l)V^2 \qquad (4.7)$$

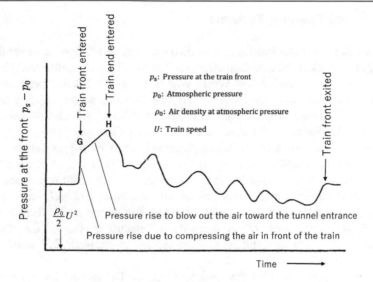

Fig. 4.72 General pattern of pressure changes in front of a train (same as Fig. 4.59)

where D_m is mechanical drag in kgf, D_a is air drag in kgf, V is train speed in km/h, W is train weight in tons, A is train cross-sectional area in m², l is train length in meters, and ρ is air density, $\frac{1}{8}$ kgs²m⁻⁴.

Figure 4.73 shows the lines calculated using the modified formula 4.7 and the measured values. The figure shows that a 12-car Shinkansen train's air drag was lower than the mechanical drag up to around 185 km/h but became higher than the mechanical drag above 185 km/h.

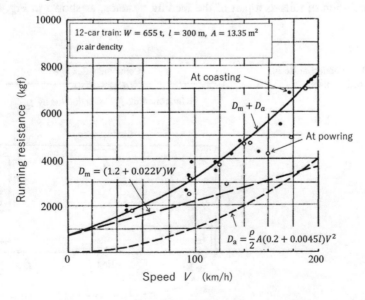

Fig. 4.73 Calculated running resistance by the modified formula. *Source* Hara, T., Ohgushi, J., Nishimura, B., "Aerodynamic Drag of Trains", Railway Technical Research Report, no. 591, RTRI, 1967, p. 5.

4.9 Power Feeding Systems

It is said that Deutsche Reichsbahn studied the first full-fledged commercial frequency railway electrification (hereafter referred to as "AC electrification") in 1935. However, due to the outbreak of World War II, it was not put into practical use at that time. After the war, Société Nationale des Chemins de fer Français (SNCF), which was in charge of Germany's railways, established AC electrification technology in 1951 based on the research conducted by Germany.

In Japan, the then JNR president Nagasaki heard of the usefulness of AC electrification from SNCF when he visited France in 1953 and started the "Alternating Current Electrification Research Committee" after returning. Since conditions such as climate, topography, social infrastructure, and technical standards differed from France, JNR set up a 23.9-km test line with a 20-kV feeding system and built two electric locomotives, one with AC electric motors and the other with DC electric motors with rectifiers, and began tests to establish JNR standards for AC electrification.

Technical issues included the choice of AC or DC motors for main motors, reducing inductive disturbance to communication lines, decreasing adverse effect on three-phase power systems, establishing noise-resistant signaling systems, and taking safety measures against short-circuit and ground fault accidents in feeding systems. Having solved these problems, JNR opened the first AC-electrified line of a total length of 28.7 km in September 1957. By the time the Shinkansen project started, the total length of AC-electrified lines had reached nearly 100 km.

The advantage of AC electrification is that high voltage can provide greater power to trains while reducing substations along lines. However, AC electrification causes inductive interference in the communication lines along railways. This is due to the inclusion of rails as a part of the feeding systems, as shown in Fig. 4.74.

Fig. 4.74 Induction noise in communication lines

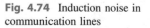

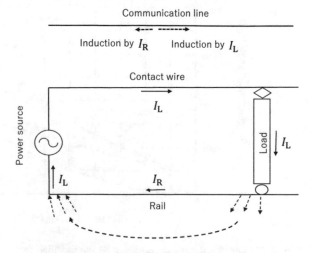

The electric current flowing through overhead wires powers the vehicle's main motors and then leaks out from rails to the ground on its way back to substations. Therefore, as shown in the figure, the noise induced in the communication line by the current I_L in the feeder line and by the current I_R in the rail does not completely cancel out, leaving noise in the communication line. Hence, in AC electrification, measures are taken to reduce the current leaking from the rails to the ground. One such measure is the use of booster transformers (BT) in feeding circuits. The AC electrification of conventional railways just described used the BT feeding system. Since there had been no problem with them, it was adopted for the Shinkansen as well.

The following is a description of the BT system's problem on the Shinkansen, the arc countermeasure.

(1) Principles of the BT Feeding System

Figure 4.75 shows a BT in the circuit that conducts electricity from the contact wire to the load (train), where Z_N and Z_R are the impedances of the negative feeder (NF) and rail.

There are two paths to return the current I_L from the load to the substation (power source in the figure), one is the NF-BT side and the other is the rail side. However, as I_L flows through the BT's primary winding, the return path must be the secondary winding of the transformer due to the transformer's electromagnetic effect. Meanwhile, for I_L to flow through the BT's primary winding, the circuit must have a break in the contact wire to insert the BT (from now on referred to as BT section), as shown in the figure. Figure 4.76 shows what happens to current distribution when a pantograph short-circuits the BT section.

In this case, the I_L flowing through the contact wire is divided into I_S and I_N (the current through the shorted section and that of BT's primary winding). Because

Fig. 4.75 Operating principle of the BT feeding system

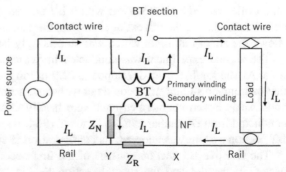

BT: Booster transformer   NF: Negative feeder

Z_N: Impedance of the route X-BT-Y Z_R: Rail impedance from X to Y

Fig. 4.76 Current distribution when the BT section is shorted

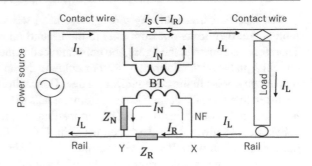

of the section short-circuited, one might think that all the I_L flows to the section side, but since I_N $(= I_L Z_R / (Z_N + Z_R))$ flows in the transformer's secondary winding, the same amount of current flows to the primary winding. Then, when the BT section in the short-circuited state is opened by the traveling pantograph, $I_S (= I_R)$ will be cut off, and an arc will appear.

(2) Arc Problems in BT Section

In May 1961, while preparations for the BT feeding system construction were underway on the Tokaido Shinkansen, it was found that a large arc had occurred on a conventional line when a long freight train passed through a BT section and damaged the overhead wires. The load current at the time was 240 A. BT section's arcing was a known issue, but it was not a problem on conventional lines because load currents were small. However, since the load currents of Shinkansen trains are much higher, about 1,000 A, it was expected that arcs would severely damage the overhead wires and pantographs, so urgent countermeasures became necessary.

In December 1962, the first test was conducted to investigate the relationship between the magnitude of the cutoff current and the arcing by using a pantograph on a trolley pulled by a motor car with 6 kV power on the overhead wire. The test showed that the arc did not extinguish when the current was over 400 A. Photo 4.26 shows the arc at that time, which was surprisingly large.

The second experiment was conducted in September 1963 using a Shinkansen train. At 100 km/h, the arc developed to 12.9 m long, and engineers were concerned that pantograph sliding strips could not withstand a single round trip between Tokyo and Osaka. Seven countermeasures were hastily considered, and eight tests were conducted from December 1962 to October 1964. As a result, the so-called resistive BT section system was adopted; its configuration is shown in Fig. 4.77.

The resistive BT section consists of the first section S_1 with a resistors R connected in parallel and the second section S_2 located 25 m away from S_1. The function of this resistive BT section is as described in the following paragraph.

Photo 4.26 Arcs observed in a simulated test (speed: 40 km/h, pantograph traveled from the left to right, interrupting current: 462 A)[16]

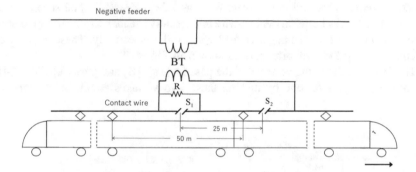

Fig. 4.77 Resistive BT section

At the time of opening, the Shinkansen train consisted of six independent two-car units, with one pantograph per unit, so the interval between pantographs was 50 m. The resistive BT section allows only one pantograph to enter between S_1 and S_2, so the current interrupted at S_1 is limited to the current of one pantograph, 240 A at most, thus preventing large arcing at S_1. Meanwhile, the pantographs that have already passed S_2 take power through BT, but when a subsequent pantograph short-circuits S_2, they receive current through both BT and R. But since the current through R is limited by R, no significant arcing occurs at S_2. The twisted section in Fig. 4.24 was needed to make up two sections at a 25-m interval and thereby prevent large arcing at the BT section.

How, then, should the value of R be determined? Masami Hayashi[17] analyzed the current distribution in this section using the schematic shown in Fig. 4.78 to determine the optimal value of R.

---

[16] *Studies of the Tokaido Shinkansen*, vol. 5, RTRI, 1964, p. 591.

[17] Masami Hayashi joined the Imperial Japanese Army in 1943, started work as RTRI in 1945 as a senior researcher, and later became head of the electric power laboratory of RTRI.

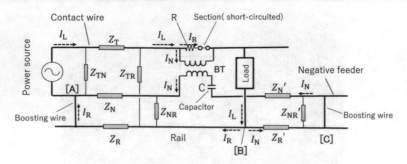

Fig. 4.78 Current flow with BT section short-circuited

In the figure, Z_T, Z_N, Z_R, and so on are the self-impedances of the contact wire, negative feeder wire, and rail, respectively; and Z_{TN}, Z_{TR}, Z_{NR}, and so on are the mutual impedances between contact wire and negative feeder wire, between contact wire and rail, and between negative feeder and rail, respectively. These values were calculated using the wire arrangement shown in Fig. 4.79.

Hayashi calculated the voltages of the pass [A]-rail-[B] and pass [A]-BT-[C]-[B] as shown in Fig. 4.78, and by making these two voltages equal, he obtained the

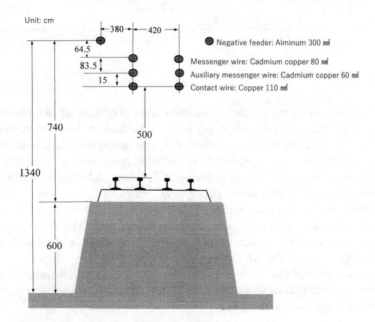

Fig. 4.79 Wire arrangement. *Source* Studies of the Tokaido Shinkansen, vol. 5, RTRI, 1964, p. 596

relationship between I_R and the constants shown in Fig. 4.78. Through this analysis, he derived that when the value of R is set to 10 Ω, I_R will be decreased to 5% of I_L.

In this way, the arcing problem of the BT section was solved and the Shinkansen started operation as scheduled. However, the resistive BT section, which required the twisted section with a complex structure, created major weakness in the catenary equipment, as described in Sect. 4.2.

4.9.1 Switch from BT Feeding System to AT Feeding System

Shortly after the Shinkansen opened for business, JNR realized that the composite compound catenary equipment with the twisted section was vulnerable to damage from trains with large number of pantographs. Therefore, JNR began researching the auto-transformer (AT) feeding system, which does not require the twisted section.

When the Shinkansen project started, the AT feeding system was discussed, but engineers adopted the BT system because the BT system had been used on conventional lines for two years without problems and because the analysis of the complex current flow of the AT system was difficult (computers were not available at the time).

Figure 4.80 shows the current flow in the AT feeder system; unlike the BT system, there is no need to have sections in the contact wire dedicated to inserting ATs. When the primary winding feeds I_2 to the train, I_1 flows through the secondary winding to cancel out the magnetic flux generated by I_2; since I_4 and I_3 have the same relationship to I_2 and I_1, the resulting current flowing through the train is $2(I_1 + I_3)$. As shown in the figure, the AT system reduces electromagnetic induction interference by not sending current to the rail in section A, where there is no train. In actual installations, the contact wire, feeder wire, and rail have self-impedance, mutual impedance, and stray capacitance, respectively. In addition,

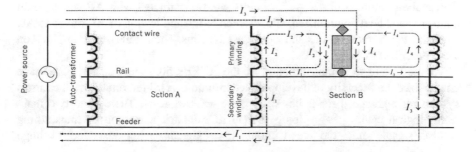

Fig. 4.80 Current distribution in the AT feeder system

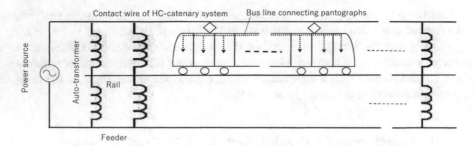

Fig. 4.81 Improved power supply system

these conductors are connected to the AT and a wire called the protection wire, thus forming complex multi-wire networks and making manual calculation of the network's current flow difficult. Therefore, the characteristic analysis of the AT system was performed after 1965 using a Bendix G-20 computer.

The AT feeding system was adopted in 1972 on the Sanyo Shinkansen, which runs westward from 'Shin-Osaka' station. This system made it possible to reduce the number of pantographs by connecting them with a high-voltage cable on the roof.

Eight years after the first opening, although having had taken a detour, the Shinkansen finally reached a stable power supply system, which used trains with a small number of pantographs, the AT feeding system, and high-tensile catenary equipment, as shown in Fig. 4.81. On the Tokaido Shinkansen, the BT feeding system was replaced by the AT system in 1991.

4.10 Signaling Safety and Route Control

4.10.1 ATC (Automatic Train Control System)

Kawanabe postulated that if track circuits were constructed with AF waves, train control would be revolutionized. His research led to application of AF waves to AC electrification's track circuits and cab signal systems. The details of this application were described in Sect. 1.4.

At the lecture in commemoration of the RTRI's 50th anniversary, Kawanabe emphasized the necessity of developing a continuous ATC (automatic train control) system for high-speed trains based on these achievements. There were two main technological challenges in doing so: (1) to establish a method of transmitting control signals onto the car without these signals being lost due to huge disturbances on the rails; and (2) to carry out this transmission safely and stably with electronic devices using a large number of diodes and transistors.

The following sections discuss how these challenges were overcome.

4.10.1.1 Invention of a Power-Synchronous SSB-type AF Track Circuit

The approximate relationship among a signal wave, carrier wave, and noise in a frequency domain is shown in Fig. 4.82. Amplitude modulation of a f_c Hz carrier wave by a f_s Hz signal wave generates three waves: $f_c - f_s, f_c$, and $f_c + f_s$. Here $f_c - f_s$ and $f_c + f_s$ are called the lower and upper sideband waves. The method of transmitting these three waves is called the double-sideband modulation method (DSB for short). In contrast, the method that transmits only one sideband waves is called the single-sideband modulation method (SSB for short).

When five signal waves are needed to control train speed, the DSB needs at least a bandwidth of 100 Hz to transmit these waves. In this case, at least one of the power harmonics comes into this bandwidth, which can make it a challenge for the signal information to be demodulated correctly.

In March 1960, a test was conducted on a conventional AC-electrified line to see if the DSB method could send five signal waves without being disturbed by noise. As expected, the ATC system using the DSB malfunctioned due to power harmonic noise. At this point, the Shinkansen ATC system ran into a snag. Akira Yusa,[18] who proposed a clever method named the power-synchronous SSB that was the key to realizing the Shinkansen ATC, wrote in his memoirs: [16]

> I felt sick to my stomach when it became impossible for the DSB method to work, and I wondered what to do about it. I thought about it for about three months, and then I came up with an idea of using a harmonic wave of the train power current as the carrier wave. (Author's translation of the Japanese).

Figure 4.83 shows the relationship between harmonic noise waves and sideband waves in the power-synchronous SSB method.

In the SSB, since the transmitter does not send the carrier wave, signal demodulation requires the receiver to create the carrier wave of exactly the same frequency as the original carrier wave, but doing so is difficult in general. However, when the same power supply is used on the transmitter and receiver side, this problem can be solved by using a frequency of harmonics of that power supply as the carrier wave frequency. In this case, even if the harmonics' frequency moves as the power supply frequency fluctuates, the range from the sideband to the adjacent harmonic wave does not change because the carrier frequency moves together with the adjacent harmonic. Therefore, the power-synchronous SSB can stably keep a bandwidth of 50 Hz (equivalent to 100 Hz in the DSB). (The name of the power-synchronous SSB comes from the fact that a common power source synchronizes both carriers.)

---

[18] Akira Yusa joined JNR in 1957. He later became director of the Electrical Facility Construction Bureau of Kyushu region after serving as a senior researcher at RTRI.

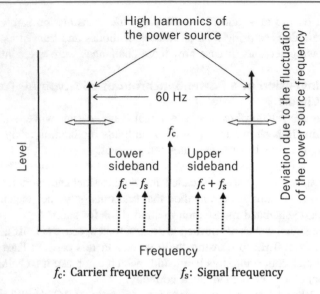

Fig. 4.82 Relationship between sideband waves and harmonic noise in the DSB

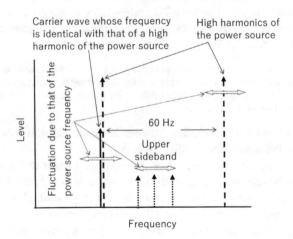

Fig. 4.83 Sideband waves and noise waves in the power-synchronous SSB system

Comparison tests between the power-synchronous SSB and DSB systems were conducted in 1960 and 1961 on an AC-electrified conventional line and confirmed that the former could withstand disturbance ten times greater than the signal voltage. This led Shinkansen's ATC to become a reality.

The power-synchronous SSB system uses some frequencies of power harmonics as the carrier wave frequency, so it does not malfunction with power harmonics. However, asynchronous harmonics can cause the system to malfunction. Therefore, asynchronous disturbance was investigated twice over the entire length of the line, and it was confirmed that there was no problem in the functioning of the ATC.

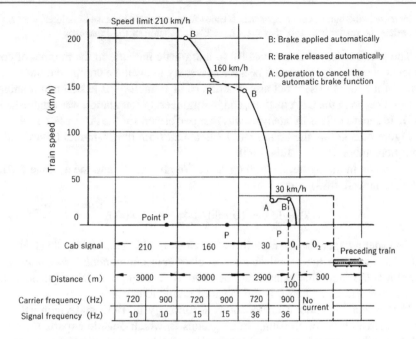

Fig. 4.84 Speed control pattern of the ATC system for the Shinkansen at its opening. *Source* Research on High-Speed Railways, Kenyusha Foundation, 1967, p. 528

Figure 4.84 shows the distance between the preceding and following train, on-board signal indication, carrier and signal waves' frequencies, and the change in train speed at the time of the opening.

The inbound track circuit used 720 Hz and 900 Hz as the carrier frequencies (the twelfth and fifteenth harmonics of 60 Hz) and upper sideband waves. In comparison, the outbound track circuit used 840 Hz and 1020 Hz as the carrier frequencies (the fourteenth and seventeenth harmonics of 60 Hz) and lower sideband waves.

4.10.1.2   Stability and Reliability of ATC System

The AF track circuit, which was first used on the Hokuriku Line in 1957, was designed to be fail-safe, as described in Sect. 1.4. However, a new problem arose because many electronic components, including vacuum tubes, were used in it. Yusa said the following [16]:

Fail-safe design was easy to achieve when the number of components was small. However, electronic signal systems have complex circuitry and use a large number of components, so conventional fail-safe theory alone is not practical because of the high number of failures. That is, increasing the reliability of the entire system is necessary. According to fail-safe approaches, transistors, for example, should be used in AC circuits by coupling to transformers rather than used in DC circuits because transistors can lead to conduction and nonconduction failures, and it is impossible to know which failure they will take. However, the accumulation of such fail-safe methods inevitably decreases the reliability due to the increase in the number of components. If trains frequently stop because of failures, they are useless, no matter how safe they are. (Author's translation of the Japanese).

The introduction of electronics led to a dramatic increase in the number of components in signal equipment. It became necessary to evaluate equipment not only in terms of it being fail-safe but also in terms of its reliability. It is said that reliability research began in the USA in 1950, and its engineering foundation was established in 1957. In Japan, reliability analysis was first performed for the AF track circuit.

Figure 4.85 shows the failure rate λ calculated from the 35-month failure history from September 1957 to July 1960.

As shown in the figure, λ was 0.44 × 10^{-4}/h after 1959, and the average lifetime T (= the inverse of λ) was

$$T = 1/\lambda = 10,000/0.44 = 22,800 \text{ h}$$

Considering that the average life of devices of this complexity level (devices with nine vacuum tubes and about 100 electronic components) was only about 1,000 h at the time, this value could be rated as "a very good value." Incidentally, the average life of 22,800 h is 2.6 years of continuous use.

JNR began its research on reliability technology in 1959 and vigorously continued thereafter by establishing study groups to which outside experts were invited. Reliability technology played a central role in developing electronic equipment for railway signals, including ATC and CTC for the Shinkansen and electronic interlocking devices that were later developed.

In addition to the power-synchronous SSB system described previously, the Shinkansen ATC incorporated many reliability improvement technologies, such as the fail-safe drive circuit for relays, separation of transmission and reception cables of track circuits, accidental contact detection of cable cores, two-out-of-three decision systems in logic sections, centralized equipment room systems (i.e., no electronic devices were placed wayside in an effort to prevent compromise or

Fig. 4.85 Failure rate of the first AF track circuit. *Source* Reliability Survey Research Report for Electronic Signal Equipment, Signal Association of Japan, 1961, p. 39

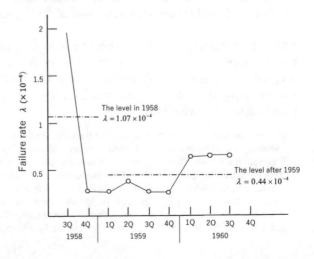

failure due to environmental conditions), and the keeping of a set of mobile equipment as a spare for a local equipment room. A systematic aging process conducted before starting ATC use also contributed to the system's stable operation after the opening.

After its opening, Shinkansen vehicles introduced power electronics technology that made them possible to increase main motor capacity and facilitate smooth acceleration/deceleration and regenerative braking. However, the ATC's environment worsened due to distorted power currents, but thanks to the power-synchronous SSB system and its high disturbance capability, the Shinkansen's safety did not waver.

A malfunction of the ATC due to asynchronous disturbance first occurred in September 1974, 10 years after the opening. Since it happened at a depot, it did not cause an accident. Still, JNR took the matter seriously as a problem that could shake Shinkansen operations' safety and implemented measures to prevent recurrence. Thus, a new version of the power-synchronous SSB system that combined two power-synchronous SSB signals to create a single piece of information was developed to solve the problem and to add new speed signals.

This system became the standard for later Shinkansen lines and was first introduced on the Tokaido and Sanyo Shinkansen from 1980 to 1993 (which coincided with the replacement of ATC system).

The power-synchronous SSB system, which had supported the Shinkansen's safety for a long time, came to the end of its mission in 1998 when digital track circuits incorporating the recent advances in digital communication technology was developed.

4.10.2 CTC (Centralized Traffic Control System)

CTC is a kind of remote-control and real-time data communication systems. It's central unit controls signaling devices such as signal lights and switch machines at stations.

CTC was developed in the USA and first installed on 64.8 km of the New York Central Railroad in 1927. In Japan, Keihin Electric Express Railway first introduced CTC in 1954 on a 4.5 km line, followed by JNR, which introduced it on a 15.7-km line with five stations in 1958. In 1963, five years later, the Shinkansen project started. At the time, the need for CTCs was not widely accepted, so there were many discussions whether to adopt CTC for the Shinkansen.

4.10.2.1 Basic Design

Conventional coding communication systems using signal relays were slow and could not be used for large-scale CTCs. Teruo Hobara,[19] the CTC developer for the

---

[19] Teruo Hobara joined the Ministry of Transport in 1945. After served as a senior researcher, he later became head of the signal laboratory of RTRI and then Director of RTRI

Shinkansen, designed this real-time control system that was orders of magnitude larger than existing CTCs as follows:

(1) Gathering Information to the Central Unit

The first step was to determine when the central unit should collect the status information of station devices (signal lights, switch machines, etc.). Because of slow data transmission speed, conventional systems using relays gathered information only when station devices changed their states. Since electronic equipment allowed for high-speed transmission, Hobara adopted a scanning method that allowed the central unit to know the devices' states at all times.

(2) Information Transmission Rate

Hobara then considered necessary transmission speed. The system must collect a large amount of information, about 2,000 bits of information in total from all stations. When a transmission error occurred, it must be corrected in the next transmission, so if the allowed time to correct the error is one second, a speed of 2000 bits per second is required. However, considering possible transmission bandwidth on a single communication line, Hobara set a development target of 1,000 bits/second (2,000 bauds) using the tri-level code format shown in Fig. 4.86.

At the time, a modulation speed of only about 120 bauds was used for electronic transmission in the world, so a speed of 2,000 bauds was literally an order of magnitude higher than the existing speed.

(3) Code Format

Since the information exchanged between the central unit and each station unit is a bit sequence of 1s and 0s, each unit must synchronize and view the bit sequence simultaneously to read the information correctly. Although there are several methods for achieving this purpose, Hobara adopted a three-level coding method suitable for the Shinkansen CTC, as shown in Fig. 4.86, in which the first half of each bit has 1 or 0 signal and the second half has no signal (i.e., each bit has a pause).

(4) System Operation Sequence

Figure 4.87 shows the procedure by which the central unit displays information about the station devices. The display cycle starts with a station signal issued from the central unit to the downstream line.

Receiving the station signal from the central unit, each station unit adds the value of its counter by one. When the counter's number is 1, the first station is selected. Then the first station unit sends out devices' information to the upstream line, and the central unit displays it on the control panel.

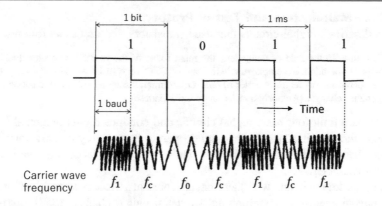

Fig. 4.86 Bit and baud used for the CTC

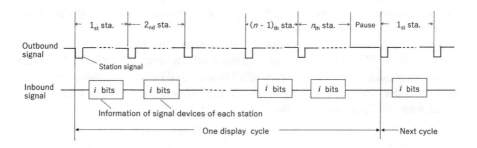

Fig. 4.87 Order of operations. *Source* Research on High-Speed Railways, Kenyusha Foundation, 1967, p. 551

Since the number of bits to be sent out from the station unit is predetermined (*i* bits in Fig. 4.87), when the central unit has received that number of bits, it reissues the next station signal.

Then the central unit and each station unit set each counter to 2, and the second station is selected and the process is repeated. When the central unit has received information from the last station (the n_{th} station in Fig. 4.87), it does not issue the station signal but enters a fixed period of pause, during which time all the counters are reset to zero. When the pause period expires, the next display cycle begins. This sequence is repeated. When the central unit issues a control command to station units, the system enters a control cycle at the end of the pause period, and then the central unit issues a signal to notify station units that the control cycle begins. Receiving this signal, each station unit sets its state to receive control information. Next, the central unit sends a station designation code. Receiving the control information, the designated station unit performs the indicated operation.

Based on the code transmission test results, some changes were made in the practical system, but the basic order of operations remained unchanged.

4.10.2.2 Manufacture and Test of Prototype

Hobara described digital circuits that used transistors and diodes as follows: [17]

> At that time, in the field of computers, the main focus of the research was on achieving
> high-speed operation. In comparison, in the case of CTC, high speed was not necessary, but
> stable operation was required under adverse environments such as voltage fluctuation and
> temperature change. (Author's translation of the Japanese).

So Hobara built four basic digital circuits and conducted environmental tests to determine the circuit constants that could operate most stably against noise, temperature change, and voltage fluctuation while taking into account the variation in transistor characteristics.

Using the logic circuits with these circuit constants, Hobara started hand-making prototype equipment (one central unit and two station units) in January 1959 (Photo 4.27).

The prototype was completed in June 1959 and continued to operate for the next two years. During this time, only five of the 836 transistors and two of the 3,862 diodes failed, both in their initial defects. Considering the subsequent improvements in transistor production technology, the prospects for the reliability of transistor logic circuits seemed bright.

Prototype production and testing of high-speed data transmission equipment began in 1959. The equipment was refurbished three times by 1961. Performance tests were conducted using cable transmission lines and microwave lines, and it was confirmed that impulsive noise was unavoidable to some extent, but transmission errors were within a practical range.

Photo 4.27 Handmade CTC experimental equipment[20]

4.10.2.3 Practical System

Through the aforementioned process, the practical system was completed as follows:

(1) **Basic Structure**

To reduce transmission errors due to pulse noise, the following changes were made in the practical system:

(i) A sequence was added so that a station unit sends back an answer to the central unit when it received control information.
(ii) Station unit designation was changed from the sequential increase of counter value method to a station code method.
(iii) The control cycle could interrupt the display cycle.
(iv) Redundancy was added to the bit structure of display signals and control signals.
(v) The system was divided into four subsystems, as shown in Fig. 4.88.

(2) **Reliability Improvement**

In order to increase system reliability, the following measures were taken:

(i) Microwave lines were allotted to backup coaxial cable lines.
(ii) Code transmission equipment was provided with a spare set so that in case of failure the faulty equipment could be switched immediately to the spare.

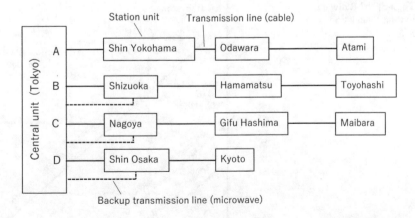

Fig. 4.88 CTC configuration at the time of Shinkansen opening. *Source* Research on High-Speed Railways, Kenyusha Foundation, 1967, p. 565

(iii)   Logic units were made triplex and operated synchronously to obtain a majority decision since, in duplex units, it is difficult to detect which unit has malfunctioned.

(iv)   Power supply was also equipped with a backup unit.

4.10.2.4 Predicted Reliability and Actual Reliability

Like the ATC, the CTC's reliability was predicted prior to its use. However, in Japan at the time there was no data about component failure rates which were the basis of prediction, so prediction was based on the failure rates published in 1961 in the USA. As a result, the system was expected to have 17 failures per month, but it had only nine failures in the first 13 months of use (0.7 failures per month), which was well below the expected rate.

Regarding these values, Hohara said the following [17]:

> The failure rate of the system was much lower than predicted because the main logic part was made triplex and the code transmission part was duplex, so most of the system failures were due to single system failures. The failure prediction was made for a system with 610,000 components, including 60,000 transistors and 250,000 diodes, but the complexity of the practical system was less than that, with 460,000 components (35,000 transistors, 170,000 diodes, 210,000 resistors, 35,000 capacitors, and 4,000 relays), so the actual failure rate should be rated as 1/15 of the predicted value. However, this value is undoubtedly at an extremely high, as can be seen from the figure [Fig. 4.89] (Author's translation of the Japanese).

Fig. 4.89 Reliability of the CTC system. *Source* Research on High-Speed Railways, Kenyusha Foundation, 1967, p. 576

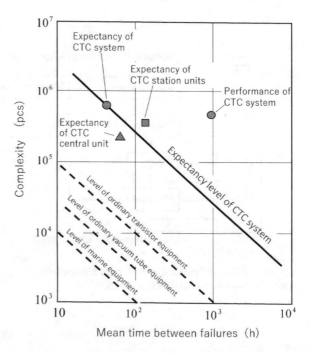

Photo 4.28 The Emperor and Empress being briefed on ATC by Matsudaira, with Kawanabe in the background (October 12, 1965). (Provided by Saburo Kiyosawa)

The aforementioned ATC and CTC were the first large-scale electronic systems for which reliability prediction was conducted in Japan, which has significance in the history of such technology. CTC was integrated with an operation control system COMTRAC, which was realized when the Sanyo Shinkansen started its operation in 1972. Since then, COMTRAC has enabled the Shinkansen's stable and high-density operation (see Sect. 6.4.1).

References

1. Tadashi Matsudaira, "Hunting Motion Tests on a Real-Size Experimental Bogie and Shinkansen Test Car's Bogies (in Japanese)", *Research on High-Speed Railways*, Kenyusha Foundation, 1967, p. 286.
2. Tadashi Matsudaira, "Effect of Elasticity of Axle Box Suspension on Hunting Critical Speed (in Japanese)", *Studies of the Tokaido Shinkansen*, vol. 3, Railway Technical Research Institute (RTRI), 1962, p. 250.
3. Tadashi Matsudaira, "Effect of Resistive Moment of Bogie Rotation on Hunting Critical Speed (in Japanese)", *Studies of the Tokaido Shinkansen*, vol. 3, RTRI, 1962, p. 254.
4. Tadashi Matsudaira, "Hunting Calculation of a Railcar with Two Bogies (in Japanese)", *Studies of the Tokaido Shinkansen*, vol. 4, RTRI, 1963, p. 221.
5. Tadashi Matsudaira, "Hunting Motion of Two-Bogie Railcar (in Japanese)", *Research on High-Speed Railways*, Kenyusha Foundation, 1967, p. 291.
6. Tadashi Matsudaira, "Derailment due to Lateral Impact Force of a Wheel (in Japanese)", *Studies of the Tokaido Shinkansen*, vol. 3, RTRI, 1962, p. 235.

7. *Ikuro Kumezawa,* "Catenary Equipment Structure for High-Speed Power Collection (in Japanese)", *Studies of the Tokaido Shinkansen,* vol. 2, RTRI, 1961, pp. 26,27
8. Masaru Iwase, *À la Carte Power Collection Technology* (in Japanese), Kenyusha foundation, 1998, p. 114.
9. [9] Hiroshi Arimoto, *From Kodama to Hikari* (in Japanese), Osaka Dengyou Co., Ltd., 1976. P. 50.
10. Kentaro Matsubara, *The Shinkansen Track* (in Japanese), Japan Railway Civil Engineering Association, 1969. P. 25.
11. Tadashi Matsudaira, "A Retrospective of Research and Development on the Tokaido Shinkansen (in Japanese)," *Journal of the Japan Society of Mechanical Engineers,* no. 646, 1972.
12. Masaharu Kunieda, "Theory and Experiment on Vertical Vibration of Rolling stock with Air Spring (in Japanese)," *Railway Technical Reseaerch Report,* no. 6, RTRI, 1958.
13. Masaharu kunieda, "Dynamic Characteristics of Air Springs for Railcars (in Japanese)", in *Studies of the Tokaido Shinkansen,* vol. 1, RTRI, 1960, pp. 116–123.
14. Tadanao Miki, "Aerodynamic Problems of High Speed Trains (in Japanese)," *The Engineering Journal for Transportation,* no. 113, Kotsu Kyouryokukai, 1955, p. 30.
15. Tomoshige Hara, "Air Pressure Fluctuation in Tunnels (in Japanese)", *Research on High-Speed Railways,* Kenyusha Foundation, 1967, pp. 360–361.
16. Akira Yusa, "Shinkansen ATC (in Japanese)", *Railway Signaling Technologies Were Born in This Way,* Japan Railway Electrical Engineering Association, 2009, p. 48.
17. Teruo Hobara, "CTC (in Japanese)", *Research on High-Speed Railways,* Kenyusha Foundation, 1967, p. 552.

Shinkansen Business Opening and Afterward

5

On October 1, 1964, the Tokaido Shinkansen opened for business. The ribbon-cutting ceremony was conducted by Reisuke Ishida, JNR president at the time. His predecessor Sogo, who had led the way in bringing the Shinkansen to fruition, had resigned 15 months earlier because the budget for the Shinkansen construction had doubled from its original amount. Sogo accepted responsibility for failing to stay within the budget (Photo 5.1).

5.1 Post-Opening Breakdowns

After opening, the Shinkansen faced a number of hardships that were not anticipated during the planning stage, including transportation disruptions due to snow, speed restrictions due to weak power supply networks, and environmental problems.

As described in Chap. 3, the decision to build the Tokaido Shinkansen was made in 1958. The technical standards and routes were finalized three years later, in August 1961. Parallel to the ongoing technical development, the rush construction work of the 515-km-long railway with many tunnels and bridges began. The final stage of technological development—performing test runs—began in mid-1962 on a 32-km-long test section and lasted until just before the opening on October 1, 1964, ten days before the Tokyo Olympics.

Construction of the Tokaido Shinkansen ran on a very tight schedule, and the opening occurred while equipment finishing and adjusting was ongoing. The embankment and roadbed of the track were still in the process of stabilization. Furthermore, the trial run before the opening revealed that the train caused voltage fluctuations in the power supply network, resulting in uneven weaving at textile factories along the Shinkansen, so deceleration operation was forced in that area. For these reasons, for the first 12 months, JNR lowered the maximum speed to 200 km/h and set the traveling time between Tokyo and Osaka to 4 h for the

© The Author(s), under exclusive license to Springer Nature Singapore Pte Ltd. 2022
T. Shimomae, *Birth of the Shinkansen*,
https://doi.org/10.1007/978-981-16-6538-7_5

Photo 5.1 Ribbon-cutting ceremony for the Shinkansen opening, October 1, 1964. (Provided by RTRI)

superexpress train Hikari (which means light) and 5 h for express train Kodama (which means echo), each of which was one hour longer than initially envisioned.

Breakdowns due to snow were frequent during the first winter of operation and resulted in many train delays and broken windows. At first, JNR couldn't understand why such things happened on snowy sections. Eventually, JNR found that the train blew up the snow on the track into the air, which adhered to car under floors in thick layers, fell in clumps, and caused ballast to scatter and damage the equipment mounted on the under floors. In some cases, the blown-up snow could get inside electrical equipment and cause a short circuit. The scattered ballast also caused the breakage of window glass. Both of these issues had not been predicted for train travel on the 32-km-long test section.

Figure 5.1 shows the number of annual breakdowns that caused train delays of 10 min or more per million kilometers of train mileage in the first five years of operation. Records at the time reported the confusing situation for six months after the opening [1]:

In the six months after the opening, the number of breakdowns per million train kilometers was 38.98, which exceeded the average of 32.77 on conventional lines. Breakdowns due to initial failures and natural disasters such as snowfall and rainfall brought about considerable train delays. Looking at the monthly changes, the number of initial failures was the highest in October. On December 18, a catenary equipment failure occurred between Shin-Yokohama and Odawara, resulting in a 14-hour suspension of service. After that, in January and February, train operation was greatly affected by snow, but it finally recovered in March. (Author's translation of the Japanese)

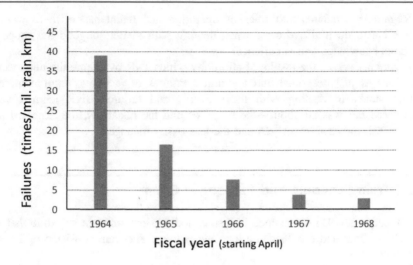

Fig. 5.1 Failures after the Shinkansen opening

The catenary equipment failure mentioned in the preceding extract is the one described in Sect. 4.2.

The Shinkansen is a complex system consisting of the following; failure in any one of these areas could result in system failure:

(1) Facilities and equipment.
(2) The many people who operate and maintain the facilities and equipment.
(3) Standards and rules that determine how to operate and maintain the facilities and equipment.

Facilities and equipment include civil engineering infrastructure (embankments, viaducts, bridges, tunnels, roadbeds, etc.), tracks, vehicles, power supply systems (power transmission and substation equipment, overhead catenary equipment), signal systems, operation control systems, communication systems, passenger information systems, and ticket sales equipment.

The test runs on the test section only confirmed that the Shinkansen vehicles could run safely at over 200 km/h. At the time of opening, JNR did not well understand what kind of damage would be caused by environmental changes such as rainfall, snowfall, strong wind, intense heat, extreme cold, and the accumulation of stress due to daily commercial operation.

The people who would be tasked with operating and maintaining the facilities and equipment, except for the very few who had experienced the work on the test section, had no idea what a high speed of over 200 km/h was like. They experienced it for the first time in pre-opening test runs.

Regarding standards and rules of operation and maintenance, there was no choice but to do trial and error when dealing with matters beyond the scope of previous knowledge and experience.

Figure 5.1 shows the results of efforts by which JNR overcame the unexpected situations as just described and the initial failures of facilities, equipment, and rolling stock. In dealing with these events and failures, JNR learned what high-speed rail was all about—the positives and the negatives, the expected and unexpected, the easily managed and the less easily managed.

5.2   Train Schedule and Passenger Count

The Tokaido Shinkansen opened in a so-called 1–1 pattern train schedule that ran one Hikari train and one Kodama train every hour. The train consisted of 12 cars with 987 seats.

With the roadbed stabilized and emergency measures against voltage fluctuations completed, JNR raised the maximum speed to 210 km/h in November 1965, 12 months after the opening, thereby shortening the traveling time between Tokyo and Osaka to 3 h 10 min for Hikari and 4 h for Kodama. JNR also reorganized the train schedule to a 2–2 pattern with two Hikari and two Kodama trains per hour, which improved convenience and significantly increased passenger numbers. In October 1967, two and a half years after the opening, the train schedule became a 3–3 pattern, with Hikari departing Tokyo at 00, 20, and 40 min and Kodama at 05, 25, and 45 min every hour.

As the number of passengers continued to increase year by year, as shown in Fig. 5.2, it became clear that the Shinkansen business, which required a huge investment, was successful. In October 1969, the train schedule became 3–6 pattern with 202 trains operated daily. For the Osaka Expo held in 1970, JNR increased the number of cars of the Hikari train to 16 cars and ran many trains, contributing to the Expo's success. Approximately 64 million people visited the Expo, of which 10 million used the Shinkansen.

5.3   Presentation at International Conference

The research on high-speed railways in Japan mentioned in Chap. 1 was unknown to the railway-advanced countries in Europe and the USA because Japan was in the post-war reconstruction period and had not yet had the opportunity to make presentations at international conferences. Matsudaira's study of railcars' hunting problem was first introduced to the world when he responded to the call from the Office for Research and Experiment (ORE) in 1955 for papers on railcar vibration. ORE is an affiliated UIC organization (Union Internationale des Chemins de Fer).

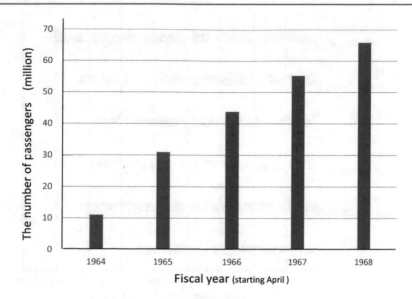

Fig. 5.2 Yearly increases in the number of passengers

Matsudaira's paper won a prize, but the prize money went to JNR, which in return sent him on a tour to Europe in 1957. During the tour he exchanged views with engineers of England, Germany, Switzerland, and Italy. According to Matsudaira's memoirs, when he told them that Japan was working on a high-speed railway with a maximum speed of 200 to 250 km/h, they replied that it was impossible and that the maximum practical speed limitation of railways would be about 140 km/h at best. It was the view of European countries at the time that test runs at high speeds were possible, but practical operation at 200 km/h would not be realistic.

Figure 5.3 shows the maximum speed of the world's railways in 1964 when the Shinkansen opened for business operation.

The second opportunity for international presentation came in July 1963, when Matsudaira participated in an ORE research committee on wheel–rail interaction. At that time, running tests on the test section had begun, and the maximum speed of 256 km/h had been reached, so Matsudaira used slides to explain the tests. The majority of people responded with admiration and encouragement. Nonetheless, some questioned the realization of the Shinkansen, saying that this was something that could be done because it was a test, and commercial operation would not go well.

The third opportunity came in November 1965, when Matsudaira participated in an international conference titled the Interaction between Vehicle and Track Convention organized by the Institution of Mechanical Engineers. Matsudaira presented a summary of a series of hunting studies as described in Sect. 4.1, including the severe hunting behavior that occurred on the test section.

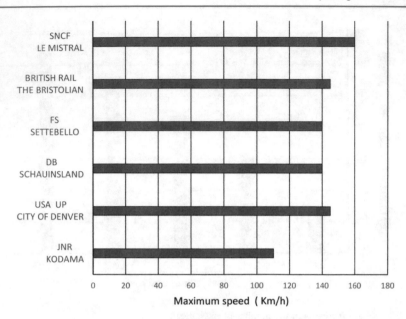

Fig. 5.3 World railway speed at the time of Shinkansen opening

The fourth opportunity came when Matsudaira participated in a symposium titled High Speed Railways and Other Forms of Guided High Speed Surface Transports held at UIC in June 1968. It had been nearly four years since the Tokaido Shinkansen had begun operation, and the initial breakdowns had finally come to an end. At the symposium, Matsudaira chaired the Rolling Stock Engineering Committee. Japanese participants presented studies titled "Aerodynamic Problems in High-Speed Trains," "Bogies of Tokaido Shinkansen Trains and Their Usage Record," and "Problems and Maintenance of Track in High-Speed Operation," all of which were highly regarded. Matsudaira wrote in his report of this symposium that five or six years ago, the term *high speed* meant less than 160 km/h, but the advent of the Shinkansen had changed the meaning of *high speed* to include speeds of 200 to 250 km/h [2].

5.4   Award

The Shinkansen received the following awards for pioneering the world of high-speed rails:

1. Christopher Columbus Award, 1966.
2. The Elmer A. Sperry Award, 1967.
3. The James Watt International Medal, to Hideo Shima, 1969.

In addition, in 2000, Shinkansen was selected as ASME's Historic Mechanical Engineering Landmarks #211. The reason for the selection was as follows: "In 1964, Shinkansen (which means "new trunk line" and is also known as the bullet train) between Tokyo and Shin-Osaka became the world's first high-speed railway system, running at a maximum business speed of over 200 km/h (130–160mph)."

References

1. *Ten Years of the Shinkansen* (in Japanese), JNR Shinkansen Office, 1975, pp. 339–340.
2. Tadashi Matsudaira, "Report on Participation in Overseas Symposium (in Japanese)," *RTRI Research Materials,* no. 9, RTRI, 1968.

Part II
The Current Shinkansen

Before discussing the current Shinkansen story, we must discuss the Japanese National Railways reform of 1987. The Shinkansen as it now exists is closely related to the JNR reform.

JNR was established in June 1959 as a spin-off of the Ministry of Transport to be self-supporting public company. It was profitable from its inception, but the development of automobiles and airplanes pushed the company into the red beginning in 1964, when the Tokaido Shinkansen began operating. Although the Shinkansen and the conventional lines in Tokyo and Osaka were in the black, the deficit on regional lines was large, so the overall deficit continued to increase every year. In 1987, to stem this deficit, the government decided to privatize JNR and split it into several companies. The three JR companies that operate in the mainland and

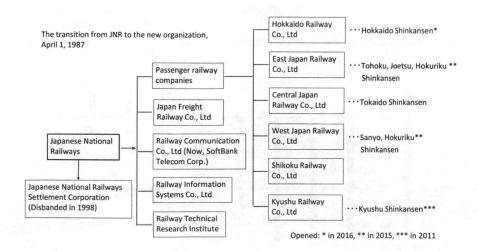

Fig 1 Transition from JNR to new organizations

the JR freight company took over a part of the JNR's debt, which had reached $353 billion in US dollars. The government took on much of the huge debt and repaid it with proceeds from the sale of land relinquished by JNR and the future proceeds from the sale of JR shares; the remainder of the debt would be repaid via taxes on the public.

Despite strong opposition from opposite parties, the JNR Reform Law was passed in November 1986. In April 1987, JNR was divided into six passenger companies, a freight company, a telecommunication company, an information systems company, and RTRI, as shown in Fig. 1. The number of personnel went from 277,000 to 201,000, as shown in Fig. 2, and decreased to 120,000 by 2019. Figure 3 shows the change in the current account balance. As can be seen from the

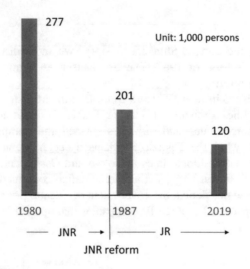

Fig 2 Changes in the number of personnel from JNR to JRs.

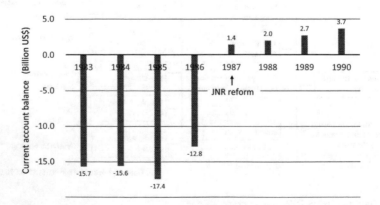

Fig 3 Improvements in current account balances

figure, JNR had run huge deficits every year, but after the reforms, the three JR companies located in the mainland started operating in the black. They started to strengthen their railways' competitiveness based on their respective strategies, including increasing the speed of the Shinkansen.

The JNR reform was a life-changing event for many people and, at the same time, a transformation that revived the Japanese railway.

The Shinkansen in 2020

6

As the number of trains increased, environmental problems such as noise, ground vibration, and interference with television signals became serious along the Shinkansen line. In June 1965, shortly after the business opening, the first complaints about the Shinkansen's noise were made to a municipal office along the line. In October 1970, some residents began a campaign for nonpayment of television subscriptions because of television disruption due to the Shinkansen. In March 1974, 575 residents filed a lawsuit demanding noise abatement and compensation. In light of this situation, in July 1975 the government issued a statement titled *Environmental Quality Standards for Shinkansen Noise*; these standards were probably more stringent than those of other countries. The aforementioned lawsuit was not settled until April 1986, when the plaintiffs and JNR entered into a settlement agreement that included noise reduction, payment of the settlement, and implementation of noise and ground vibration control work. The JNR reforms took place the following year, and three JRs in the mainland began to develop technology to improve speed while meeting environmental requirements so the Shinkansen would be competitive with airplanes and automobiles.

Table 6.1 shows the transitions of the Shinkansen cars. During the 20 years when JNR was the sole operator, there were only three types of Shinkansen cars, including the Series 0 at the time of the Shinkansen's opening. After JNR changed into JRs, 14 new types of vehicles were designed and placed into service in 33 years. The stagnation during the JNR era can be attributed to the financial problems of the JNR, as mentioned previously.

Table 6.2 shows the technologies that have been studied to date since the opening of the Tokaido Shinkansen. They cover a wide range of technologies, including safety improvements such as earthquake countermeasures and ATC improvements; snow countermeasures that enabled operation of the Joetsu and Tohoku Shinkansen in extreme winter weather; and many technological studies that achieved environmental protection and speed improvement.

Fortunately, TV interference has been solved by the digitalization of TV broadcasting, which started in 2003.

© The Author(s), under exclusive license to Springer Nature Singapore Pte Ltd. 2022   217
T. Shimomae, *Birth of the Shinkansen*,
https://doi.org/10.1007/978-981-16-6538-7_6

Table 6.1 Changes in Shinkansen vehicles

Series	Company	Power Frequency	1960s	1970s	1980s	1990s	2000s	2010s	2020s	Route	Maximum Speed (km/h)
Series 0	JNR	60	1964				2008			Tokaido/Sanyo	210
Series 200	JNR	50			1982			2013		Tohoku/Joetsu	240
Series 100	JNR	60		Partly double decker	1985			2012		Tokaido/Sanyo	220/230
series 400	JR East	50				1992		2010		Yamagata	240/130*
Series 300	JR Central	60				1992		2012		Tokaido/Sanyo	270
Series E1	JR East	50	Double decker			1994		2012		Tohoku/Joetsu	240
Series E2	JR East	50/60				1997				Hokuriku/Tohoku/Joetsu	260/275/240
SeriesE3	JR East	50				1997				Akita/Yamagata	275/130*
Series 500	JR West	60				1997				Tokaido/Sanyo	270/300
Series E4	JR East	50	Double decker			1997			2021	Tohoku/Joetsu	240
Series 700	JR Central/West	60				1999			2020	Tokaido/Sanyo	270/285
Series 800	JR Kyushu	60					2004			Kyushu	260
Series N700	JR Central/West	60					2007			Tokaido/Sanyo/Kyushu	285/300/260
Series E5	JR East	50						2011		Tohoku/Hokkaido	320
Series E6	JR East	50						2013		Akita	320/130*
Series E7	JR East/West	50/60						2014		Hokuriku/Joetsu	260/240
Series N700S	JR Central	60						2020		Tokaido/Sanyo	285/300

JNR ——— 1987 ——— JR companies

*On conventional lines

Table 6.2 Technological progress after the opening of the Tokaido Shinkansen

Technology	Effect									
	High Safety	High Top Speeds	High Average speeds	High Punctuality	High-Frequency operation	High Riding quality	Low Noise emissions	Low Ground vibrations	Energy Saving	Labor Saving/Low Cost/Long Life
Early earthquqke detection and alarm system	○									
Safety measures against big earthquakes	○									
Dual frequency combination ATC	○									
Train radio system using leaky coaxial cable				○	○					
Digital ATC	○		○	○	○	○			○	
Downsized highpower motors driven by power electronic devices		○						○	○	○
Bolsterless bogies		○						○	○	○
Ruducing of wheel diameters		○						○	○	
Arc wheel profile		○								
Reducing of car cross-sections		○					○	○	○	

(continued)

Table 6.2 (continued)

Technology	High Safety	High Top Speeds	High Average speeds	High Punctuality	High-Frequency operation	High Riding quality	Low Noise emissions	Low Ground vibrations	Energy Saving	Labor Saving/Low Cost/Long Life
Improved nose shape of leading cars		O					O		O	
Car-body tilting at curved sections			O			O			O	
Improvement of track maintenance standards						O				O
Hollow extruded alminum alloy for car bodies		O				O		O	O	
Active suspension						O				
Inteligent traffic control systems				O	O					
Anti-yaw dampers between cars						O				
High tensile strength contact wire		O								
Countermeasures against snow				O						
Slab track				O		O				O
Rail grinding							O	O		O
Ballast mats							O	O		

(continued)

Table 6.2 (continued)

Technology	Effect									
	High Safety	High Top Speeds	High Average speeds	High Punctuality	High-Frequency operation	High Riding quality	Low Noise emissions	Low Ground vibrations	Energy Saving	Labor Saving/Low Cost/Long Life
Sleepers with elastic pads							○	○		
Reducing the number of pantographs							○			○
Cover-all hood of train couplers							○		○	
Low noise pantographs							○			
Side barriers for pantographs and insulators							○			
Wayside noise barriers							○			
Stabilizer for power supply systems using power electronics technology				○	○				○	
Regenerative brake systems		○				○		○	○	○
High-speed inspection train for the electrical equipment and the track				○						○

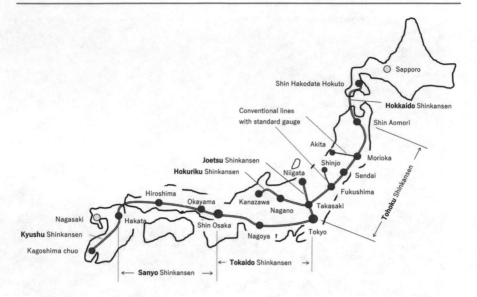

Fig. 6.1 Shinkansen lines as of 2020

The Shinkansen system is characterized by an excellent safety record, high speed, environmental friendliness, high punctuality, and high-density operation. The following is an overview of the system.

6.1 Shinkansen Lines and Maximum Speed

Figure 6.1 shows the Shinkansen lines as of 2020. There are seven lines: Tokaido, Sanyo, Tohoku, Joetsu, Hokuriku, Kyushu, and Hokkaido. The total length of these lines is about 3,000 km, and the maximum speeds are as follows: 320 km/h in the Tohoku Shinkansen, 300 km/h in the Sanyo Shinkansen, and 285 km/h in the Tokaido Shinkansen. The Tohoku Shinkansen is currently undergoing trials to increase its maximum speed to 360 km/h so that it can compete with airlines and offer customers another high-speed travel option.

6.2 Train Frequency, Travel Time, and Punctuality of the Tokaido Shinkansen

Table 6.3 shows the Tokaido Shinkansen's outbound timetables in 1964, 1990, and 2019 from 8:00 a.m. to 9:00 a.m. In the table, H, K, and N stand for Hikari, Kodama, and Nozomi trains, respectively.

Table 6.3 Tokaido Shinkansen timetables from 8:00 to 9:00 a.m. in 1964, 1990, and 2019.

1964 : 1H-1K Pattern

km		H1 (Hikari)	(Kodama) K1	H1
0	Tokyo	800	830	900
29	Shin-Yokohama	·	849	·
84	Odawara	·	915	·
105	Atami	·	928	·
180	Shizuoka	·	1005	·
257	Hamamatsu	·	1047	·
294	Toyohashi	·	1109	·
366	Nagoya	1029	1145	1129
396	Gifu-Hashima	·	1202	·
446	Maibara	·	1229	·
514	Kyoto	1134	1302	1234
553	Shin-Osaka	1200	1330	1300
Travel time between Tokyo and Osaka		4h.	5h.	

1990 : 5H-2K Pattern

km		H1	H2	(Kodama) K1	(Hikari) H3	H4	K2	H5	H1
0	Tokyo	800	804	816	824	840	848	856	900
29	Shin-Yokohama	·	821	833	·	857	905	·	·
84	Odawara	·	·	855	·	·	930	·	·
105	Atami	·	·	905	·	·	940	·	·
121	Mishima**	·	·	915	907	·	952	·	·
180	Shizuoka	·	·	946	·	·	1022	·	·
257	Hamamatsu	·	·	1021	·	·	1058	·	·
294	Toyohashi	·	·	1041	1003	·	1116	·	·
366	Nagoya	954	1003	1116	1027	1038	1149	1046	1054
396	Gifu-Hashima	·	·	1131	·	·	1205	·	·
446	Maibara	·	·	1154	·	1103	1226	·	·
514	Kyoto	1040	1048	1220	1112	1128	1252	1132	1140
553	Shin-Osaka	1056	1104	1236	1128	1144	1308	1148	1156
Travel time between Tokyo and Osaka		2h.56min.		4h.					

2019 : 10N-2H-2K Pattern

km		N1	H1	N2	N3*	N4*	N5*	(Kodama) K1	N6	(Hikari) H2	N7*	N8*	N9	N10*	K2	N1
0	Tokyo	800	803	810	813	820	823	826	830	733	840	847	850	853	856	900
7	Shinagawa**	807	810	817	820	827	830	834	837	740	847	854	857	900	904	907
29	Shin-Yokohama	819	721	829	832	839	842	845	849	752	859	806	909	912	916	919
84	Odawara	·	·	·	·	·	·	903	·	808	·	·	·	·	935	·
105	Atami	·	·	·	·	·	·	912	·	·	·	·	·	·	944	·
121	Mishima**	·	·	·	·	·	·	924	·	·	·	·	·	·	957	·
146	Shin-Fuji**	·	·	·	·	·	·	939	·	·	·	·	·	·	1012	·
180	Shizuoka	·	911	·	·	·	·	955	·	·	·	·	·	·	1026	·
229	Kakegawa**	·	·	·	·	·	·	1010	·	·	·	·	·	·	1041	·
257	Hamamatsu	·	937	·	·	·	·	1022	·	·	·	·	·	·	1055	·
294	Toyohashi	·	·	·	·	·	·	1041	·	·	·	·	·	·	1113	·
336	Mikawa-anjo**	·	·	·	·	·	·	1058	·	·	·	·	·	·	1152	·
366	Nagoya	941	1007	947	954	959	1004	1109	1011	917	1019	1028	1031	1036	1143	1041
396	Gifu-Hashima	·	·	·	·	·	·	·	·	930	·	·	·	·	1157	·
446	Maibara	·	·	·	·	·	·	·	·	952	·	·	·	·	1219	·
514	Kyoto	1019	1046	1024	1033	1036	1043		1049	1013	1057	1127	1110	1116	1240	1119
553	Shin-Osaka	1033	1100	1037	1046	1050	1056		1103	1026	1110	1120	1123	1130	1253	1133
Travel time between Tokyo and Osaka					2h.27min.					2h.53min.					3h.57min.	

* Nonregular service

** Station built after the opening

The Nozomi train, which began operation in 1992 using Series 300 cars, was the first Shinkansen train to run at a speed of 270 km/h while achieving a high level of environment protection.

As shown in Table 6.3 and Fig. 6.2, the train frequency of the Tokaido Shinkansen has increased dramatically since the opening. Lately, an average of more than 370 trains per day operate with virtually no delays. Precisely, including delays due to natural disasters such as typhoons, earthquakes, and heavy snow, the average delay was 33 s per train for the four years from 2016 to 2019. As the number of trains increases, one train's delay delays all subsequent trains, which makes it difficult to restore on-time operation. Therefore, maintaining punctuality becomes much more challenging as the number of trains increases. The fact that 370 trains are operated on time every day shows the high degree of perfection of all elements of the Shinkansen, including facilities, equipment, vehicles, train operation management system, maintenance work, and work management systems.

The operation of up to 370 trains per day is managed by a computer system called COMTRAC (short for Computer-aided Traffic Control system). High-frequency operation and punctuality of the Shinkansen owes much to this system, which is one of the two major achievements that have improved the Shinkansen's quality. The other is the safety measures against major earthquakes. These two improvements will be discussed later.

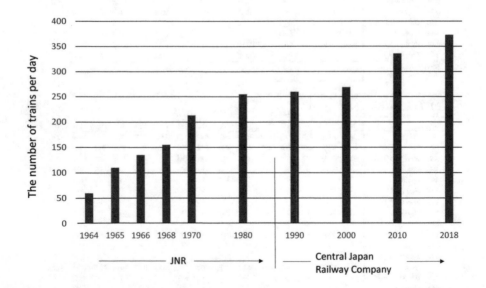

Fig. 6.2 Number of trains on the Tokaido Shinkansen

6.3 Safety

When people talk about high-speed railways, they tend to focus on maximum speed, but safety is essential for railways. As described in Sects. 4.1 and 4.2, the barriers to improving speed were clarified at the stage of the Shinkansen's realization. A roller rig to simulate high-speed running was developed at the time. Nowadays, advanced analysis of vehicle dynamics is possible by combining a high-performance computer and a high-speed roller rig that can generate track irregularity, give the vehicle running resistance, and so on. It is now possible to increase maximum speeds as long as the track is almost straight. At a high-speed symposium held by the UIC in 1968, four years after the Shinkansen went into service, a speed of 350 km/h was considered to be realistically possible. Even speeds up to 400 km/h were discussed. However, as a matter of course, higher speed must go hand in hand with higher safety. The higher the speed, the more disastrous the accident will be by leaps and bounds.

6.3.1 High-Speed Rail Accidents

It is not possible to quantify how potentially unsafe a railway is before an accident occurs. This is because the inherent riskiness of a railway only becomes apparent when an accident occurs.

Table 6.4 shows high-speed rail accidents worldwide (these data were retrieved from an Internet search). Some accidents did not result in death, but the author judged them to be serious enough to include in this table. There could have been more accidents that were serious but were not included in the table. Please note that the information compiled in the table is subject to partial errors because it was sourced from the Internet.

In addition to the accidents listed in the table, according to data retrieved from the Internet, the TGV experienced eight level-crossing accidents in the 20 years from 1988 to the end of 2007. Of the eight, six derailed, and four resulted in fatalities; two, fortunately, did not derail, but they all must have been dangerous accidents. It is presumed that in addition to these accidents, a certain number of accidents would have occurred in the railways that have level crossings with roads.

In the column for the cause of each accident in Table 6.4, the author has noted the cause judging from the information on the Internet.

6.3.2 Safety of the Shinkansen

Currently, the number of Shinkansen cars is about 4,600. Shinkansen trains which are organized with these cars travel about 440,000 km per day, the distance of 38 round trips between Paris and New York, or 11 orbits of the Earth; 200 round trips to the moon, or 4,000 laps of the Earth in one year. One may be surprised at these

Table 6.4 High-speed rail accidents

NO	Date	Country	Train	Accident Type[a]	Cause[b]	Deaths	Speed (Relative speed) in km/h	Annotation
1	12/14/1992	France	TGV	1	A	0	240	Wheel sticking
2	12/21/1993	France	TGV	1	N	0	294	Track sinking
3	1/12/1997	Italy	ETR460	2	H	8	158	The train ran at 158 km/h on a 105 km/h limit section
4	9/19/1997	UK	HST	3 → 1, 4	H	7	140	Ignoring a red signal, the train entered a prohibited section, collided with a freight train, and burned. ATS maintenance was defective
5	6/3/1998	Germany	ICE	2	A, H	101	200	Wheel's tire breakage. Insufficient consideration of tire strength
6	10/5/1999	UK	HST	3	H	31	143(209)	The train entered a prohibited section without checking the red signal at the place with poor visibility
7	6/5/2000	France	Eurostar	1	A	0	250	Locomotive driving device damaged
8	10/17/2000	UK	IC225	1	A, H	4	185	Rail breakage on a conventional line
9	2/28/2001	UK	IC225	1	A	10	201	The train collided with a car that had fallen on the track, derailed, and collided with an oncoming freight train
10	11/17/2001	Germany	ICE	The train ran into a wrong track	H	0	180	A signal cable was misconnected. The train entered a 80 km/h limit section at 180 km/h. The engineer took prompt action to avoid the danger
11	10/31/2001	France	TGV	1	A	0	130	Rail breakage on a conventional line

(continued)

Table 6.4 (continued)

NO	Date	Country	Train	Accident Type[a]	Cause[b]	Deaths	Speed (Relative speed) in km/h	Annotation
12	10/17/2001	France	Eurostar	4	A	0		A broken cardan joint shaft caused a fire in the lead car
13	10/23/2004	Japan	Shinkansen	1	N	0	200	Derailment due to a direct earthquake
14	11/3/2007	Korea	KTX	3 → 2	H	0	30	An ATS alarm sounded, but the driver failed to brake and crashed
15	7/1/2011	China	CRH	3 → 2	A,H	40		Signal system defects and train dispatcher's improper instruction after the failure
16	7/24/2013	Spain	Alvia	1 → 2	H	78	190	The train entered a 80 km/h limit section at 190 km/h
17	7/17/2014	France	TGV	3	H	0	30(60)	An express train collided with the rear end of a TGV running at about 30 km/h at about 90 km/h
18	11/14/2015	France	TGV	1 → 2	H	11	265	The train ran at 265 km/h on a 176 km/h limit section
19	12/8/2018	Korea	KTX	1	H	0	Slow speed	Incorrect connection of a signal cable caused the train to enter a wrong track
20	12/13/2018	Turky	YHT	3	H	9		The train collided with a track maintenance vehicle due to the lack of a signal system
21	2/6/2020	Italy	ETR400	1 → 3	H	2	180	
22	3/5/2020	France	TGV	1	N	0	270	Landslide along the line

[a] Accident type 1: Derailment 2: Overturn 3: Collision 4: Fire
[b] Cause A: System/Parts defects H: Human errors N: Natural disasters.

numbers, but the Tokaido Shinkansen alone runs 175,000 km a day (350 trains × 500 km). There is no doubt that the cumulative train kilometers to date since the Tokaido Shinkansen opening in 1964 is astronomical. The number of passengers on the Tokaido and Sanyo Shinkansen has exceeded 9.1 billion since the Tokaido Shinkansen opening. Including other routes, the number exceeds 13.3 billion, which means that everyone in the world has taken the Shinkansen twice.

The feat of the Shinkansen is that it opened up the world of high-speed railways and has carried 13.3 billion people, partly due to good fortune, without a single fatality in 56 years. The following is an overview of factors that contributed to the very high safety of the Shinkansen.

(1) Design to Maintain Safety

(i)   The Shinkansen operates on dedicated lines that do not share tracks with conventional lines. The importance of this is illustrated by the fact that, worldwide, more than a few accidents have been caused by the sharing of tracks with conventional lines. The sharing of tracks entails a certain amount of danger.

(ii) Shinkansen has no road-level crossings. Accidents due to road-level crossings cannot be prevented by railway's efforts alone since they involve vehicles driven by ordinary people. Road-level crossings are highly dangerous facilities for high-speed rails (Photo 6.1).

Photo 6.1 Road overpass of Shinkansen. (Provided by akira/PIXTA)

(iii) To prevent cars and other objects from falling onto Shinkansen tracks, substantial protective barriers are installed at locations where this risk is a factor. In addition, detection devices are linked to the ATC in case of a fall.
(iv) All areas are fenced to prevent animals and people from entering the track.
(v) Central command monitors situations on the track via anemometers and rain gauges installed along the line. When the measured value exceeds specified figures, central command issues a slow-down order. (Safety measures against earthquakes are described later.)
(vi) Intervals between trains are controlled by the ATC, which is characterized by fail-safe logic and high reliability.
(vii) When the ATC is not available, trains are operated manually based on the prescribed rule at speeds up to 110 km/h. Drills are carried out regularly.

(2) **Legislation to Maintain Safety**

Interfering with the Shinkansen operation or standing on Shinkansen track is strictly prohibited by the government's *Special Act for Punishing Conduct that Interferes with Shinkansen Safety.*

(3) **Management to Maintain Safety**
Railways operate and maintain their facilities, equipment, and vehicles daily. As noted above, even if they are designed, fabricated, and constructed with all possible care, safety cannot be maintained if the operation and maintenance are shoddy.

Heinrich's law (Fig. 6.3) states that there are 29 less serious accidents and 300 minor irregularities behind every serious accident. This ratio will vary depending on the subject matter, but the concept seems to apply to railway accidents.

The ability to avoid serious accidents depends on being sensitive to the 29 nonserious accidents and 300 minor abnormalities represented in the figure. If we ignore these minor accidents and abnormalities, there is a high probability that a major accident will eventually occur. Small irregularities are the voice of heaven

Fig. 6.3 Heinrich's law

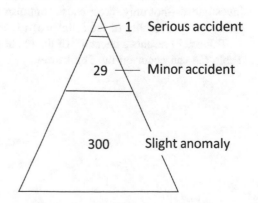

Table 6.5 Classification of accident factors

Category		Rules/Standards		Human		
		No Rules R1	Insufficient R2	Negligence H1	Insufficient Ability H2	Inappropriate Working Conditions H3
Design	D	D-R1	D-R2	D-H1	D-H2	D-H3
Construction/Production	C	C-R1	C-R2	C-H1	C-H2	C-H3
Operation	O	O-R1	O-R2	O-H1	O-H2	O-H3
Maintenance	M	M-R1	M-R2	M-H1	M-H2	M-H3
Natural disaster	N					

that tells us of potential dangers of which we are unaware. Therefore, the investigation of accidents/incidents/abnormalities causes must be logical, thorough, and consider all precedents and possible contributing factors. Generally, accidents/incidents/abnormalities can be attributed to problems in the design, fabrication/construction, or operation/maintenance stages. Since a person performs the work in each stage according to prescribed rules, if there is a problem with the result, the cause is either the person or the rules. In the case of the person, the cause is either negligence of duty, lack of ability, or unreasonable working conditions. Table 6.5 shows the classification of accident factors. The key to preventing accidents is to develop personnel who have the necessary knowledge and correct skills to complete their work with a sense of responsibility; to create an appropriate work environment; and to have a system in place for checking and fairly evaluating work performance. If there are flaws in the rules, of course, they must be corrected. Earthquakes and other natural disasters fall outside the aforementioned prevention strategies.

In the past, the Shinkansen experienced incidents, as shown in Table 6.6. Fortunately, these did not result in significant accidents but were just one step away from having tragic results. The causes were thoroughly investigated and made public, along with countermeasures and background factors. No similar events have occurred since then.

For the Shinkansen, the causes of even minor abnormalities are thoroughly investigated—not only direct causes but also underlying factors. The high level of safety and punctuality of the Shinkansen is the result of such efforts.

Table 6.7 presents a checklist of the likelihood of whether the accidents listed in Table 6.4 can occur on the Shinkansen.

Table 6.6 Incidents on the Shinkansen

Date	Status	Cause	Countermeasures	Underlying Factors
4/25/1966	An axle broke while driving	Product defect	Improvement of quality control	C-R2
2/21/1973	ATC issued a stop signal to a deadhead train joining the main line from a depot, but the train did not stop and entered the main line	The oil applied to rail sides adhered to rail tread surfaces and caused wheels to slide	Improvement of work manuals	O-R1
9/12/1974	ATC issued an incorrect signal in a depot	Malfunction due to noise	Change of system configuration	D-R2
9/30/1991	Wheels stuck and treads were severely damaged	Pruduct defect	Improvement of quality control	C-R2, O-H2
5/6/1992	Three of the four bolts that fixed a traction motor to the bogie frame fell off, damaging the gearbox	Incorrect work	Improvement of work manuals	M-R2
6/27/1999	A concrete block fell from a tunnel ceiling	Incorrect work	Improvement of work quality	C-H1
10/2/2004	Derailment due to a near-field eathquake	Natural disaster	Reinforcement of civil infrastructures and vehicle improvement	–
12/11/2017	Cracks occured in a bogie frame	Pruduct defect	Improvement of quality control	C-R2, O-H2

In preparing to write this section, the author surveyed (via the Internet) the world's high-speed rail accidents and was surprised to find that there were so many serious accidents. In addition to the accidents listed in the table, there must surely hve been many more incidents that have not been made public but might have resulted in serious consequences. Tables 6.4 and 6.7 show the excellent safety of the Shinkansen.

Table 6.7 Possibility of accidents on the Shinkansen

NO	Train	Accident type[a]	Cause[b]	Annotation	Possibility on the Sinkansen	Rationale
1	TGV	1	A	Wheel sticking	No	All axles have slide detecting device with high reliability
2	TGV	1	N	Track sinking	No	There is no such a possible place
3	ETR460	2	H	The train ran at 158 km/h on the 105 km/h limit section	No	All lines are protected by ATC
4	HST	3 → 1, 4	H	Ignoring a red signal, the train entered a prohibited section, collided with the freight train, and burned. ATS maintenance was defective	No	All lines are protected by ATC
5	ICE	2	A, H	Wheel's tire breakage. Insufficient consideration of tire strength	No	No special wheels are used
6	HST	3	H	The train entered a prohibited section without checking the red signal at the place with poor visibility	No	All lines are protected by ATC
7	Eurostar	1	A	Locomotive driving device damaged	Yes	Damage to some parts cannot be denied
8	IC225	1	A, H	Rail breakage on a conventional line	No	Shinkansen lines do not share tracks with conventional lines
9	IC225	1	A	The train collided with a car that had fallen on the track, derailed, and collided with an oncoming freight train	No	Every potential place has strong protective measures
10	ICE	The train ran into a wrong track	H	A signal cable was misconnected. The train entered a 80 km/h limit section at 180 km/h. The engineer took prompt action to avoid the danger	No	Separate specialists check the design and construction of signal devices. Once construction is done, the function is checked by another specialist
11	TGV	1	A	Rail breakage on a conventional line	No	Shinkansen lines do not share tracks with conventional lines

(continued)

Table 6.7 (continued)

NO	Train	Accident type[a]	Cause[b]	Annotation	Possibility on the Sinkansen	Rationale
12	Eurostar	4	A	A broken cardan joint shaft caused a fire in the lead car	Yes	Damage to some parts cannot be denied
13	Shinkansen	1	N	Derailment due to a direct earthquake	Yes	Derailment will be possible even with the measures described in Sect. 6.4
14	KTX	3 → 2	H	An ATS alarm sounded, but the driver failed to brake and crashed	No	All lines are protected by ATC
15	CRH	3 → 2	A,H	Signal system defects and train dispatcher's improper instruction after the failure	No	The Shinkanser is equipped with high-reliability ATC designed to be fail-safe. Commanders are highly skilled
16	Alvia	1 → 2	H	The train entered a 80 km/h limit section at 190 km/h	No	All lines are protected by ATC
17	TGV	3	H	An express train collided with the rear end of a TGV running at about 30 km/h at about 90 km/h	No	All lines are protected by ATC
18	TGV	1 → 2	H	The train ran at 265 km/h on a 176 km/h limit section	No	All lines are protected by ATC
19	KTX	1	H	Incorrect connection of a signal cable caused the train to enter a wrong track	No	When signal cable connection is completed, the function is always confirmed by another expert
20	YHT	3	H	The train collided with a track maintenance vehicle due to the lack of a signal system	No	All lines are protected by ATC
21	ETR400	1 → 3			No	
22	TGV	1	N	Landslide along the line	No	There is no place for such a possibility

[a] Accident type 1: Derailment 2: Overturn 3: Collision 4: Fire

[b] Cause A: System/Parts defects H: Human errors N: Natural disasters

6.4   Two Significant Achievements After 1964

6.4.1   COMTRAC (Computer-Aided Traffic Control System)

Mamoru Hosaka, who had moved from the Japanese Navy to the RTRI in 1945, returned to Japan in 1953 after a year of study at the Massachusetts Institute of Technology. Inspired by evolving computer technologies in the USA, Hosaka foresaw the fundamental importance of introducing computerized automatic control and information processing technology to the railway.

At the time, computers in Japan were still in their infancy, so RTRI installed a Bendix G-15 computer in 1957. This computer used vacuum tubes for logic circuits and a magnetic drum for the main memory. With a new tools in hand, RTRI set about developing computer systems for railways. Although the G-15's capacity was limited, the instruction manual with schematics helped RTRI's engineers learn digital circuitry techniques. Based on this knowledge, in 1960, Hosaka and Ohno[1] hand-built a computer with 2,000 transistors and 10,000 diodes and created Japan's first online train seat reservation system, which was capable of selling 2,400 seats on four trains per day. It was the beginning of today's many useful computer systems, such as various reservation systems and online payment capabilities.

In the field of train operation control, in 1958, Hobara started to develop an electronic CTC system for the Tokaido Shinkansen, as described in Sect. 4.10. Yamamoto[2] then began to develop a Programmed Route Control System (PRC for short), which was the core of the train operation control system named COMTRAC.

COMTRAC was first used for the run from Tokyo to Okayama in 1972, when the Sanyo Shinkansen, the western extension of the Tokaido Shinkansen, opened to Okayama (see Fig. 6.1). Since then, COMTRAC has been improved many times and has acquired artificial intelligence that can suggest changing the train schedule to accommodate the numerous train delays. The current version of COMTRAC is in its ninth generation.

Figure 6.4 shows the current COMTRAC configuration, which consists of three subsystems: Electronic Data Processing (EDP for short), Programmed Route Control (PRC), and Man–Machine Advanced Processor (MAP). Roughly speaking, each subsystem works as follows [1]: The EDP creates the train schedule for each day that includes both regularly scheduled and irregularly scheduled trains and sends this schedule to the PRC, which, based on the train timetable from the EDP, instructs the CTC to create the route for each train. Upon receiving the instructions, the CTC controls the station devices to create the route for each train and sends to the MAP train locations and train number information (i.e., which train is running on which track circuit). The MAP processes this information to verify that trains are

---

[1] Yutaka Ohno joined the Ministry of Transport in 1946. After serving as a senior researcher at RTRI, he became the head of the computer systems laboratory of RTRI, and later became a professor at Kyoto University, and the chairman of the Information Processing Society of Japan.
[2] Ichiro Yamamoto joined RTRI in 1952. After served a senior researcher, he became head of the computer center of RTRI and later became a professor at Toyo University.

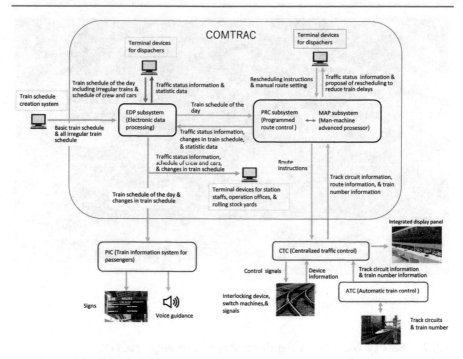

Fig. 6.4 Configuration of COMTRAC

running on time and displays all trains' status on the terminal device at the train dispatcher. When trains are delayed, the PRC creates several rescheduling proposals to recover the delays and sends them to the train dispatcher's terminal device. The train dispatcher selects the best one and sends it back to the PRC for execution.

Before COMTRAC implemented artificial intelligence of route suggestion, rescheduling was handled by the train dispatcher and was based on his experience. However, as the number of trains increased, many factors came into play and made it difficult for even the most experienced commander to create a good solution in a limited time. Fast, high-capacity computers and improved software have made the current COMTRAC possible. The current COMTRAC monitors and manages more than 400 trains a day (Photo 6.2).

Photo 6.2 Control center. (Provided by JR Central)

6.4.2 Safety Measures Against Major Earthquakes

As noted in Sect. 6.3, the Shinkansen is a very safe high-speed railway. However, it is not immune to the danger of major earthquakes. Studies on ensuring safety against earthquakes began in 1964, and the Tokaido Shinkansen soon introduced an earthquake warning system that involved installing seismographs at each substation and ensuring that substations would shut off power to the train if the seismographs detected tremors above the predetermined value. By 1991, JNR had installed 100 seismographs on the 2,000-km Shinkansen line and 300 on the 20,000 km of conventional lines.

Figure 6.5 shows the seismic waves and the alarm criteria in the earthquake warning system. This system was a major step forward in ensuring safety against earthquakes, but there were problems. If the shaking was so great that it derailed

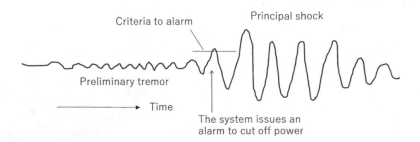

Fig. 6.5 Operation of earthquake warning system

trains after the warning, the warning was too late. On the other hand, the warning was unnecessary if the earthquake stopped soon after the warning. The Urgent Earthquake Detection and Alarm System (UrEDAS), developed by Yutaka Nakamura,[3] solved these problems.

Figure 6.6 shows the principles of UrEDAS operation. When a seismograph located far from the Shinkansen detects an earthquake's pre-shock (primary wave), the system identifies the epicenter, estimates the earthquake's magnitude, and decides whether to issue an alarm. If necessary, the system shuts off power to trains, and trains automatically apply emergency brakes to slow down before a sizeable main shock (secondary wave) reaches them. This system takes advantage of the fact that a small shaking foreshock is a longitudinal wave that travels rapidly. In contrast, a high shaking main shock is a transverse wave that travels at a slower rate than the foreshock wave. The Tokaido Shinkansen introduced this system in 1992. Many seismometers were installed on coastlines and along the Shinkansen line.

Currently, all Shinkansen lines are equipped with the latest version of UrEDAS, so if an earthquake occurs far from the train, the train can slow down before the main earthquake hits so it can stop without derailing, or, if it does derail, the derailment will not result in a catastrophe. However, if the epicenter is close to the train, the train could derail at high speed, leading to a disaster.

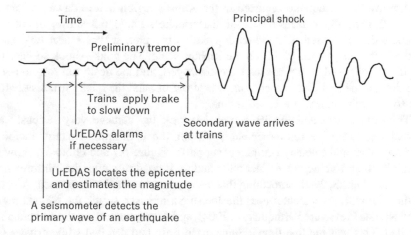

Fig. 6.6 Operation of UrEDAS

---

[3] Yutaka Nakamura joined JNR in 1978. After serving as a senior researcher at RTRI, he became the president of System and Data Research Co., Ltd.

Photo 6.3 Collapsed viaduct. (Author)

A powerful earthquake occurred in the Kansai region before dawn on January 17, 1995 (The Great Hanshin-Awaji Earthquake). Photo 6.3 shows Shinkansen viaducts that collapsed near the earthquake's epicenter. It was a near-field earthquake with a magnitude of 7.3 on the Richter scale. The quake shook so hard that buildings collapsed, many houses were destroyed, and the death toll rose to 6,434. The time of the first Shinkansen train was at 6 a.m., and the quake struck at 5:46 a. m., so the train narrowly escaped harm.

The government and the JR companies took the damage very seriously and carried out extensive reinforcement work on the civil infrastructure, including viaduct pillars and catenary equipment supports. Figure 6.7 and Photo 6.4 show the methods of reinforcing the viaduct pillar and their condition after the reinforcement.

On October 23, 2004, something that everyone was afraid of happened. A strong earthquake with an epicenter near the Joetsu Shinkansen struck the Niigata region (Mid Niigata Prefecture Earthquake in 2004), and a train running on the outbound line derailed. This was the first time a Shinkansen train had derailed while driving (see entry 13 in Table 6.4). However, thanks to the aforementioned reinforcement work, the viaduct was not damaged (Photo 6.6). The train, which was running at 200 km/h, derailed, continued traveling for 1.6 km and stopped without rolling over, so no one was injured (Photo 6.5). If the viaduct had not been reinforced, the train would have rolled over and the number of injuries and fatalities would have been catastrophic.

Also contributing to the lack of injuries and fatalities in the Niigata derailment was that there was no train coming from the inbound line. If there had been an inbound train, it could have collided with the derailed train. The Ministry of Transport and JR companies considered it merely a luck that a horrible accident did

Fig. 6.7 Reinforcement
method for viaduct pillar

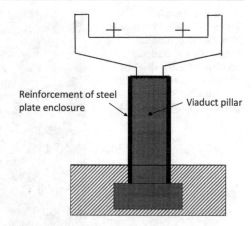

Reinforcement of steel
plate enclosure

Viaduct pillar

Photo 6.4 Viaduct pillar
after reinforcement. (Author)

not occur, and they decided to take further measures against earthquakes. They arrived at the following measures; measure (1) has been carried out since the 1955 earthquake:

(1) Reinforce civil infrastructure, including viaducts and embankments, to prevent tracks from deforming significantly in the event of an earthquake.
(2) Install guardrails to prevent train derailment (Photo 6.7).
(3) Install anti-deviation devices on the bogie frames or axle boxes of cars to ensure that trains can stop without overturning in the event of a derailment (Fig. 6.8).
(4) Strengthen emergency braking (e.g., spray fine ceramic particles between wheel and rail to prevent sliding).

Photo 6.5 Derailed
Shinkansen train. (Reprinted
from the "Report on the
Investigation of Railway
Accidents RA2007-8-1" by
Aircraft and Railway
Accidents Investigation
Commission, MLIT, p. 38)

Photo 6.6 Derailed
Shinkansen train on viaduct.
(Provided by masa9/PIXTA)

Photo 6.7 Guardrail
(Author)

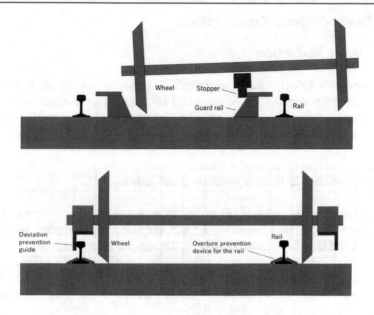

Fig. 6.8 Devices that prevent deviation from the track

The Hanshin-Awaji and Niigata earthquakes were the near-field type. In contrast, the massive earthquake that occurred at 14:46 on March 11, 2011, was an ocean-trench earthquake. (The 2011 off the Pacific coast of Tohoku Earthquake). The epicenter was located 130 km off the coast at a depth of 24 km. The magnitude measured nine on the Richter scale. The quake, which caused severe shaking and huge tsunamis that stretched about 500 km north to south, killed 15,899 people and left 2,529 missing. The damage to the railways, including the Shinkansen, was enormous, including the collapse of catenary equipment poles with insufficient seismic reinforcement. In contrast, the civil infrastructures, which had undergone seismic reinforcement, were hardly damaged. Therefore, although more than 30 Shinkansen trains were running on the Tohoku Shinkansen Line at the time of the earthquake, they stopped without derailing because the tracks remained intact and thanks to the earthquake warning system, except one test run train without passengers which derailed at a low speed.

Seismic reinforcement of civil infrastructures such as viaducts, levees, and bridges is still ongoing, but high-priority sections of Shinkansen lines have been completed. More than 68,000 viaduct pillars have been reinforced, as shown in Photo 6.4.

The Great Hanshin-Awaji Earthquake occurred 31 years after the Shinkansen service was launched. As mentioned previously, when this earthquake hit, the Shinkansen was unprotected against major earthquakes; protective measures were initiated in the aftermath of that earthquake and continue today. The author believes that these countermeasures have made catastrophes due to earthquakes mostly avoidable. This is a momentous achievement in the history of railway's safety.

6.5   Environmental Conservation

6.5.1   Noise Reduction

The loudness of sound noise is expressed in a unit of decibel A. It is a value obtained by weighting the values measured by a sound level meter with the frequency characteristics of human hearing (A characteristics) and is written as dB(A).

The noise standards for the Shinkansen issued in 1975 were as follows:

70 dB(A) or less in residential areas.
75 dB(A) or less in areas of commerce and industry.

The standard value of 70 dB(A) or 75 dB(A) is the average of the top 10 peak noise values of 20 trains measured by a A-characteristics-weighted sound level meter with the time average processing characteristics set to S (slow).

However, since each measured value is in logarithmic dB(A), the average values is not the simple average of dB values. For example, when the peak noise values of three trains are 60 dB(A), 70 dB(A), and 80 dB(A), then the average of these three, $L_{A,S\,max}$, is calculated as follows:

$$L_{A,S\,max} = 10\log_{10}\frac{1}{3}\left(10^{\frac{60}{10}} + 10^{\frac{70}{10}} + 10^{\frac{80}{10}}\right) = 10\log_{10}\frac{1}{3}\left(10^6 + 10^7 + 10^8\right)$$
$$= 75.7\ dB(A) \tag{6.1}$$

The subscripts in $L_{A,S\,max}$ indicate the value is A- characteristics- weighted, S-characteristics- weighted, and based on maximum noise value of each train.

The term *noise level* as used here is not meant in a general sense. It refers to a physical value measured, taking into account human hearing characteristics. Figure 6.9 shows the noise levels measured and calculated by the aforementioned method at a distance of 25 m from the track along the Tokaido Shinkansen [2]. The various technological developments shown in Table 6.2 have reduced the value to 75 dB(A) at a distance of 25 m from the track. As shown in the figure, the Series 300 trains, introduced in 1992, were the first to achieve this noise level at 270 km/h. Subsequently, while achieving the same or better low-noise performance as the Series 300, the JR West developed the Series 500 (maximum speed of 300 km/h) in 1997, the JR central developed the Series 700 (285 km/h) in 1999, and the JR East developed the Series E5 (320 km/h) in 2011. In the JR East's currently ongoing efforts to increase speeds to 360 km/h on the Tohoku Shinkansen, the challenge is how to reach these speeds without compromising safety and environmental protection standards.

The method for determining noise ratings in Japan is as described in the preceding discussion. However, the method for obtaining the rating value differs from country to country. Globally, many countries use the equivalent noise method to obtain the rating value. This method, denoted as L_{Aeq}, evaluates the average noise during a certain timeframe (including when trains are not running), such as daytime,

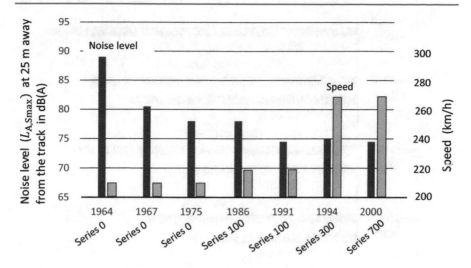

Fig. 6.9 Trends in train noise levels (no soundproof walls in 1964)

evening, or night. Hence, the value is much smaller, even if measured at the same place, than the Japanese method which only evaluates each train's peak noise level.

In the case of the Tokaido Shinkansen, a comparison of $L_{A,Smax}$ and L_{Aeq} measured at a point 25 m away from the track showed that the latter was about 15 to 18 dB(A) lower than the former. The difference will be much larger on the Tohoku Shinkansen, which has a lower train density than the Tokaido Shinkansen. As stated previously, Japan and European countries have different ways of determining the evaluation values, so the standards based on these values differ and it is not possible to use their measured values for the sake of comparison. Nonetheless, the author believes that the Shinkansen has the world's highest level of environmental protection performance.

6.5.2   Energy Conservation

The energy consumed by high-speed trains depends on many factors, including running resistance, train's weight, and braking systems. Figure 6.10 shows that the Series N700 cars introduced in 2007 consume about half as much electricity at 220 km/h as did the first Shinkansen car, the Series 0. The figure also shows that vehicles with less noise and less ground vibration have excellent energy efficiency, which means that high-speed vehicles are not sacrificing environmental protection, but rather are becoming more environmentally friendly.

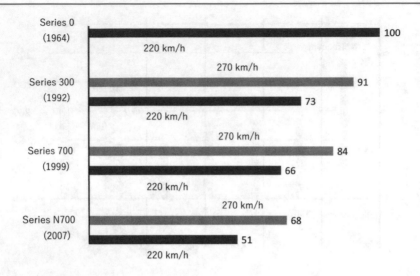

Fig. 6.10 Increased performance in energy saving

References

1. Narihito Kon, "Update of COMTRAC for Tokaido and Sanyo Shinkansen (in Japanese)", *Railway & Electrical Engineering,* no. 9, 2013, pp. 31–34.
2. Mitsuru Ikeda, "Noise and Countermeasures for Catenary Equipment-Pantograph System (in Japanese)", *Rolling Stock and Technology,* no. 96, 2004, p. 28.

Part III
Photographs

Photographs (1964–2020)

7

7.1 Series 0 (1964–2008), Tokaido and Sanyo Shinkansen

The world's first commercial high-speed vehicle. A total of 3,216 cars were manufactured with repeated improvements. Body: Steel, axle load: 16 tons, traction
motor: DC motor, electric brake system: dynamic braking, maximum speed:
210 km/h (Photos 7.1, 7.2, 7.3, 7.4, 7.5, 7.6).

Photo 7.1 Series 0 train running in city area (provided by JR Central)

© The Author(s), under exclusive license to Springer Nature Singapore Pte Ltd. 2022   247
T. Shimomae, *Birth of the Shinkansen*,
https://doi.org/10.1007/978-981-16-6538-7_7

Photo 7.2   Series 0 train running through snowed area (provided by JR Central)

Photo 7.3   Driver's cab
(provided by rolling/PIXTA)

Photo 7.4   General overhaul
(provided by yanmo/PIXTA)

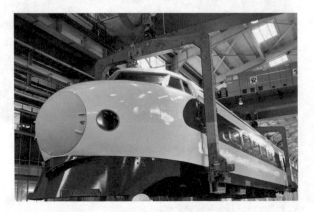

Photo 7.5   General overhaul (provided by yanmo/PIXTA)

Photo 7.6   Series 0 16-car train passing through tea plantation (provided by JR Central)

7.2 Series 200 (1982–2013), Tohoku and Joetsu Shinkansen

Cold and snow-resistant specifications for snowy areas. Equipped with a snowplow that removes snow on the track sideways without letting snow fly up. Body: Aluminum alloy, axle load: 17 tons, traction motor: DC motor, electric brake system: dynamic braking, maximum speed: 240 km/h (Photos 7.7, 7.8, 7.9).

Photo 7.7 Face of the Series 200 lead car (provided by tarousite/PIXTA)

Photo 7.8 Profile of the Series 200 lead car (provided by photo-uny/PIXTA)

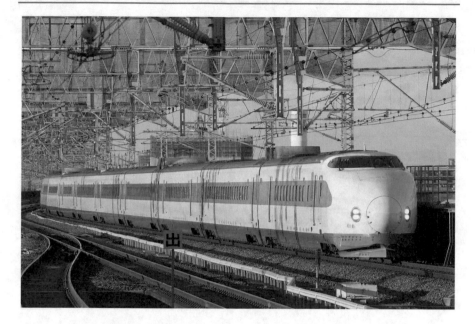

Photo 7.9 Series 200 10-car train (provided by MarineMarron/PIXTA)

7.3 Series 100 (1985–2012), Tokaido and Sanyo Shinkansen

The second-generation vehicle on the Tokaido and Sanyo Shinkansen. Partially double-decker structure. Body: Steel, axle load: 15 tons, traction motor: DC motor, electric brake system: dynamic braking and eddy current braking, maximum speed: 230 km/h (Photos 7.10, 7.11, 7.12, 7.13).

Photo 7.10 Shape of the Series 100 lead car (Author)

Photo 7.11 Series 100 16-car train (provided by JR Central)

Photo 7.12 Series 100 train leaving Tokyo station for Shin-Osaka (provided by JR Central)

Photo 7.13 Series 100 train running through snowy area (provided by JR Central)

7.4   Series 400 (1992–2010), for Yamagata via Tohoku Shinkansen

In 1992, the track of the conventional line between Fukushima and Yamagata was upgraded to the standard gauge (see Fig. 6.1). JR East built the Series 400 vehicle that operates on the Tohoku Shinkansen from Tokyo to Fukushima and on the conventional line from Fukushima to Yamagata. The vehicle's size is smaller than that of the Shinkansen to meet the conventional line's structure gauge. The maximum speed is 130 km/h on the conventional line.

Body: Steel, axle load: 13 tons, traction motor: DC motor, electric brake system: dynamic braking, maximum speed: 240/130 km/h (Photos 7.14 and 7.15).

Photo 7.14 Series 400 7-car train on the Tohoku Shinkansen (provided by OGC/PIXTA)

Photo 7.15 Series 400 train running on snowed curved section of conventional line (provided by alpha7000/PIXTA)

7.5   Series 300 (1992–2012), Tokaido and Sanyo Shinkansen

The third-generation vehicle on the Tokaido and Sanyo Shinkansen, developed by JR Central to strengthen the competitiveness of the Shinkansen. With new technologies, the performance of the Shinkansen vehicle was renewed. With this vehicle, the maximum speed on the Tokaido Shinkansen increased from 220 to 270 km/h while achieving a high-level of environment protection. The travel time between Tokyo and Shin-Osaka shortened from 2 h 49 min to 2 h 30 min. Body: Aluminum alloy, bogie: bolsterless, axle load: 11.3 tons, traction motor: AC motor, electric brake system: regenerative braking and eddy current braking, maximum speed: 270 km/h (Photos 7.16, 7.17, 7.18, 7.19).

Photo 7.16 To achieve both high speed and environmental protection, Series 300 lead car's shape has changed significantly from Series 0, 200, and 100. (Author)

Photo 7.17 Series 300 16-car train passing through Lake Hamana (provided by JR Central)

Photo 7.18 Series 300 train running through suburban area (provided by JR Central)

Photo 7.19 Series 300 train running through tea plantation (provided by JR Central)

7.6 Series E1 (1994–2012), Tohoku and Joetsu Shinkansen

A double-decker-type mass transport vehicle developed by JR East to support commuting to work and school. Body: Steel, bogie: bolsterless, axle load: 17 tons, traction motor: AC motor, electric brake system: regenerative and air supplement control braking, maximum speed 240 km/h (Photos 7.20 and 7.21).

Photo 7.20 Series E1 12-car train (provided by Kintaro/PIXTA)

Photo 7.21 Series E1 train at Tokyo station (provided by shiokake/PIXTA)

7.7 Series E2 (1997–), Tohoku, Joetsu, and Hokuriku Shinkansen

Developed by JR East. There are two types of vehicles: vehicle number of 0s and 1000s. The former is a 50/60 Hz model with strong hill-climbing power for operation on the Hokuriku Shinkansen, which straddles 50 Hz and 60 Hz areas and has a steep 30‰ gradient section. This vehicle transported visitors to the 1998 Nagano Olympics. The latter is a 50 Hz-only model for the Tohoku/Joetsu Shinkansen with active suspension to improve ride comfort. Body: Aluminum alloy, bogie: bolsterless, axle load: 13 tons in 0s and 12 tons in 1000s, traction motor: AC motor, electric brake system: regenerative braking, maximum speed: 275 km/h (Photos 7.22, 7.23, 7.24, 7.25).

Photo 7.22 Series E2 16-car train (provided by Rei/PIXTA)

Photo 7.23 Series E2 lead car's shape (Author)

Photo 7.24 Series E2 9-car train running in city area (provided by tarousite/PIXTA)

Photo 7.25 Low-noise pantograph on low-noise insulator (Author)

7.8   Series E3 (1997–), for Akita/Yamagata via Tohoku Shinkansen

In 1977, the track of the conventional line between Morioka and Akita was upgraded to standard gauge (see Photo 6.1). JR East built the Series E3 vehicle to operate on the Tohoku Shinkansen line from Tokyo to Morioka and on the conventional line from Morioka to Akita. The vehicle's size in cross section and length is smaller than that of the Shinkansen to accommodate the conventional line. The maximum speed is 130 km/h on the conventional line. Body: Aluminum alloy, bogie: bolsterless, axle load: 12 tons, traction motor: AC motor, electric brake system: regenerative braking, maximum speed: 275/130 km/h (Photos 7.26 and 7.27).

Photo 7.26 Series E3 6-car train on the Tohoku Shinkansen (provided by Seventh Heaven/PIXTA)

Photo 7.27 Series E3 train running on conventional line (provided by Seventh Heaven/PIXTA)

7.9   Series 500 (1997–), Sanyo and Tokaido Shinkansen

Developed by JR West with the aim of increasing its competitiveness to airplanes. For environmental protection, characterized by a 15-m-long nose, a slightly smaller cross-sectional area close to circular, and T-shaped pantograph. Body: Aluminum alloy, bogie: bolsterless, axle load: 11.2 tons, traction motor: AC motor, electric brake system: regenerative braking, maximum speed: 300 km/h (Photos 7.28, 7.29, 7.30).

Photo 7.28   Series 500 16-car train running on curved section (provided by T. Tsuchiya/PIXTA)

Photo 7.29   Series 500 lead car's shape (provided by Toshinori baba, CC BY-SA 3.0, via Wikimedia Commons)

Photo 7.30   Series 500 8-car train running on the Sanyo Shinkansen in autumn leaves (provided by photoAC)

7.10   Series E4 (1997–2021), Tohoku and Joetsu Shinkansen

An all-double-decker Shinkansen train developed by JR East following the Series E1. The leading car of the 8-car train is equipped with a split / merge device and can also operate as a double-connected 16-car train. The capacity of 1,634 passengers for a 16-car train is the largest in the world for high-speed trains, supporting commuting to work and school. Body: Aluminum alloy, bogie: bolsterless, axle load: 16 tons, traction motor: AC motor, electric brake system: regenerative and air supplement control braking, maximum speed: 240 km/h (Photos 7.31, 7.32, 7.33, 7.34, 7.35).

Photo 7.31   Series E4 train leaving Omiya station for Niigata (provided by F4UZR/PIXTA)

Photo 7.32 Series E4 16-car train stopping at Takasaki station (provided by tarousite/PIXTA)

Photo 7.33 Series E4 train running on snow-free tracks thanks to snow melting equipment, Joetsu Shinkansen (provided by tarousite/PIXTA)

Photo 7.34   Stairs to the second floor (Author)

Photo 7.35   Series E4 8-car train running in early autumn. (provided by K481/PIXTA)

7.11   Series 700 (1999–2020), Tokaido and Sanyo Shinkansen

The fifth-generation vehicle on the Tokaido and Sanyo Shinkansen, developed by JR Central and JR West as the replacement for Series 0 and 100. By devising the nose shape, it achieved both high speed and environmental protection with the short nose of 9.2 m. This vehicle became the technical basis for subsequent vehicles. A total of 1,328 cars were manufactured. Body: Aluminum alloy double skin, bogie: bolsterless, axle load: 10 tons, traction motor: AC motor, electric brake system: regenerative and eddy current braking, maximum speed: 285 km/h (Photos 7.36, 7.37, 7.38, 7.39, 7.40, 7.41).

Photo 7.36 Series 700 16-car train running in snowy area, Tokaido Shinkansen. (provided by YKK1/PIXTA)

Photo 7.37 Mt. Fuji and Series 700 train (provided by JR Central)

Photo 7.38 Series 700 16-car train on the Tokaido Shinkansen (provided by JR Central)

Photo 7.39   General overhaul (provided by railstar/PIXTA)

Photo 7.40   General overhaul (provided by railstar/PIXTA)

Photo 7.41   General overhaul (provided by yanmo/PIXTA)

7.12   Series 800 (2004–), Kyushu Shinkansen

This is the vehicle for the Kyushu Shinkansen first developed by JR Kyushu, 6-car train. The basic structure is the same as the Series 700, but the head shape, interior lining, seats, and equipment layout were changed. Some trains can equip with inspection devices for the track and electric equipment on the ground. Body: Aluminum alloy double skin, bogie: bolsterless, traction motor: AC motor, electric brake system: regenerative, maximum speed: 260 km/h (Photos 7.42 and 7.43).

Photo 7.42 Series 800 6-car train on the Kyushu Shinkansen (provided by Noriemon/PIXTA)

Photo 7.43 Series 800 train passing near the sea (provided by Sakura/PIXTA)

7.13   Series N700 (2007–), Tokaido, Sanyo, and Kyushu Shinkansen

The sixth-generation vehicle on the Tokaido and Sanyo Shinkansen, and the second generation on the Kyushu Shinkansen. Based on the Series 700, this vehicle was jointly developed by JR Central and JR West, achieving higher speed, comfort and improved environmental performance. Equipped with the body tilting device, the vehicle raised the speed limit of 255 km/h on the curved section of the Tokaido Shinkansen to 270 km/h. The further improved N700A shortened the traveling time with an increased curve running ability. Body: Aluminum alloy double skin, bogie: bolsterless, axle load: about 10 tons, traction motor: AC motor, electric brake system: regenerative, maximum speed: 300 km/h (Photos 7.44, 7.45, 7.46, 7.47, 7.48, 7.49, 7.50, 7.51, 7.52, 7.53).

Photo 7.44 Series N700 train stopping at Nagoya Station; the white building in the background is the headquarters of JR Central Company (provided by JR Central)

Photo 7.45 Series N700
lead car's shape (provided by
photoAC)

Photo 7.46 Series N700
train running through
snow-covered area (provided
by JR Central)

Photo 7.47 Series N700
16-car train running on curved
section (provided by JR
Central)

Photo 7.48 Driver's cab (provided by JR Central)

Photo 7.49 Green class (Author)

Photo 7.50 Standard class (Author)

Photo 7.51 Cover-all hood and anti-yawing damper between cars (Author)

Photo 7.52 Low-noise pantograph (Author)

Photo 7.53 Platform screen door, Tokyo station (Author)

7.14   Series E5 (2011–), Tohoku and Hokkaido Shinkansen

The E5 series developed by JR East is a new generation of Shinkansen vehicle that combines advanced technologies with a high level of running performance, reliability, environmental performance and comfort. To protect the environment, it has a long nose with a devised shape, low-noise pantographs and pantograph sound insulation plates, full bogie covers, and cover-all hoods at car joints. Besides, it is equipped with full active suspension and a body tilting system to provide comfortable ride at high speed. Body: Aluminum alloy double skin, bogie: bolsterless, axle load: 11.7 tons, traction motor: AC motor, electric brake system: regenerative, maximum speed: 320 km/h (Photos 7.54, 7.55, 7.56, 7.57, 7.58, 7.59, 7.60, 7.61).

Photo 7.54   Series E5 10-car train (provided by YKK1/PIXTA)

Photo 7.55   Lead car's profile (Author)

Photo 7.56 Series E5 train passing through Hachinohe City in winter. (provided by tarousite/PIXTA)

Photo 7.57 Series E5 train entering the 54 km long (23 km undersea) Seikan Tunnel which connects the mainland and Hokkaido (see Photo 6.1–1). (provided by T2/PIXTA)

Photo 7.58 Series E5 and E6 coupled train running on the Tohoku Shinkansen (provided by F4UZR/PIXTA)

Photo 7.59 Low-noise pantograph and noise insulating side pate (Author)

Photo 7.60 Cover-all hood and anti-yawing damper between cars (provided by Rsa, CC BY-SA 3.0, via Wikimedia Commons)

Photo 7.61 Standard class (Author)

7.15 Series E6 (2013–), for Akita via Tohoku Shinkansen

Series E6 was developed by JR East as the successor to Series E3, which connects Tokyo and Akita via the Tohoku Shinkansen. The cars are smaller than the Shinkansen cars to accommodate conventional line. It operates in conjunction with the Series E5 between Tokyo and Morioka. Body: Aluminum alloy double skin, bogie: bolsterless, axle load: 11.4 tons, traction motor: AC motor, electric brake system: regenerative, maximum speed: 320/130 km/h (Photos 7.62, 7.63).

Photo 7.62 Series E6 train at Tokyo station (provided by tarousite/PIXTA)

Photo 7.63 Series E6 train running on snow-covered conventional line (provided by EF5861/PIXTA)

7.16   Series E7 (2014–), Hokuriku and Joetsu Shinkansen

The Series E7 is the second generation of the Hokuriku Shinkansen, jointly developed by JR East and JR West. The system configuration is compatible for 50/60 Hz power supply and has the enhanced climbing and braking performance to accommodate the 30‰ gradient section on the Hokuriku Shinkansen. The nose length of the leading car is 9.1 m, the same as that of the Series E2. This Series has a Gran class that is superior to green cars. Body: Aluminum alloy double skin, bogie: bolsterless, traction motor: AC motor, electric brake system: regenerative, maximum speed: 260 km/h (Photos 7.64, 7.65, 7.66, 7.67, 7.68, 7.69, 7.70, 7.71).

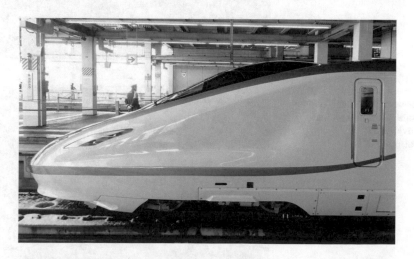

Photo 7.64 Lead car's shape (provided by Rsa, CC BY-SA 3.0, via Wikimedia Commons)

Photo 7.65 Series E7 12-car train leaving Omiya station for Tokyo (provided by Akihiko Murakami/PIXTA)

Photo 7.66 Series E7 train running through mountainous area of the Hokuriku Shinkansen (provided by Peisama/PIXTA)

Photo 7.67 Series E7 train runs with Tateyama Mountain Range in the background. (provided by JuraiJurai/PIXTA)

Photo 7.68 Gran class (Author)

Photo 7.69 Green class (Author)

Photo 7.70 High-voltage cable on the roof connecting pantographs (Author)

Photo 7.71 Low-noise
pantograph on low-noise
insulator (Author)

7.17   Series N700S (2020–), Tokaido and Sanyo Shinkansen

The seventh-generation vehicle on the Tokaido and Sanyo Shinkansen developed by JR Central, incorporating latest technologies. It is equipped with the "Battery-powered self-movable system", the first attempt at the high-speed railway. Even if power is cut off from the overhead line due to disaster, the train can move to a safe place so that passengers can evacuate. It is also equipped with fully active vibration control devices to improve riding comfort.

Body: Aluminum alloy double skin, bogie: bolsterless, traction motor: AC motor, electric brake system: regenerative, maximum speed: 300 km/h (Photos 7.72, 7.73, 7.74, 7.75, 7.76).

Photo 7.72   Series N700S 16-car train running on curved section (provided by F4UZR/PIXTA)

Photo 7.73 Series N700S train running near Atami station (provided by tarousite/PIXTA)

Photo 7.74 Series N700S train running near Shin-Fuji station (provided by nozomi/PIXTA)

Photo 7.75 Series N700S train running on curved section (provided by photoAC)

Photo 7.76 Series N700S train and rice field before harvest (provided by JR Central)

7.18   Inspection Train for the Electric Equipment and the Track

Japan's first track inspection vehicle was the YA9000, built in 1925 [1], and the first electric equipment inspection vehicle was the KUMOYA93000 (Photo 4.12), built in 1959. Inspection vehicles for the Shinkansen were built based on these technologies. First, one inspection vehicle for the track was built in 1963 followed by the electric equipment inspection vehicles named T1 that were converted from the 4-car Train B (Photo 4.7). At that time, as the track inspection vehicle used mechanical detectors that touch the rails, which could not accommodate high speeds, track inspection was performed at night at 130 km/h. After that, noncontact detectors using light were developed, which allowed performing inspection at the same speed as commercial trains. Meanwhile, with regard to electric equipment, the control of contact wire wear became extremely important because contact wires often rapidly wore locally, sometimes broke, and caused major transportation problems. However, there was no technology to measure contact wire wear from an inspection vehicle. This problem was solved by Horiki[1] who, using ITV technology, succeeded in developing a device to determine the contact wire wear from the scanning line information of contact wire's sliding surface images taken from an inspection car. The number of the scanning was 500 times per second, so at 210 km/h, the device could measure contact wire wear every 12 cm. Due to strong sunlight during the daytime, measurements could only be performed at night, but this invention was a major advance in electric equipment inspection. This device, by employing a laser beam, soon evolved to be usable in the daytime while increasing the number of measurements per second. The track inspection car and electric equipment inspection cars were combined into one 7-car inspection train called T2, in 1974, which could perform the daytime inspection at the commercial train's speed (210 km/h). The train was yellow, so it was later called a Doctor Yellow. Meanwhile, the track inspection detectors became to accommodate much higher speeds, so Doctor Yellow T4, which was built as the successor to T2 in 2001, and "East i," which inspects the Tohoku, Joetsu, and Hokuriku Shinkansen, built in 2001, became possible to perform the inspection at 275 km/h.

Currently, JR East has one "East i," and JR Central and JR West have one Doctor Yellow each, T4 and T5. These inspection trains diagnose the health conditions of the electric equipment (power supply system, catenary equipment, signaling system, and train radio system) and the track about every ten days. In recent years, inspection by commercial trains has also been progressing (Photos 7.77, 7.78, 7.79, 7.80, 7.81, 7.82, 7.83, 7.84).

---

[1] Kenji Horiki joined the Ministry of Transport in 1946. He later became a senior researcher at RTRI.

Photo 7.77 7-car Doctor Yellow T2 train (1974–2001) (provided by T. Tsuchiya/PIXTA)

Photo 7.78 7-car Doctor Yellow T4 train (2001–) (provided by JR Central)

Photo 7.79 Doctor Yellow T4 train performing inspection (provided by JR Central)

Photo 7.80 6-car East i train (2001–) (provided by KUMOYUNI/PIXTA)

Photo 7.81   East i performing inspection (provided by JonJonpe/PIXTA)

Photo 7.82 Measurement
work in the Doctor Yellow
(provided by JR Central)

Photo 7.83 Track
maintenance work (provided
by JR Central)

Photo 7.84 Catenary
equipment maintenance work
(provided by JR Central)

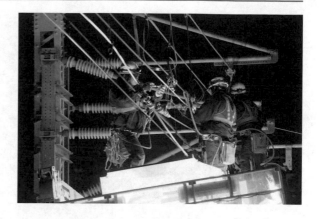

7.19   Monuments

Photo 7.85 Monument in honor of the then JNR President Sogo, built on the west end of the 18–
19 platform of Tokyo station. (Author)

Photo 7.86 Commemorative plaque at Tokyo station, reading "New Tokaido Line: Product of the wisdom and effort of the Japanese people." (Author)

Photo 7.87 Start point mark of the Shinkansen, embedded around car No.8 on the platform 18–19 of Tokyo station. (Author)

Reference

1. Yosuke Tsubokawa, "Track Inspection Technology from Vehicles (in Japanese)", *Journal of the Society of Instrument and Control Engineers*, no.2, 2007.

Chronology

Year	Day-Month	Event
1958	18-Apr	Construction Standards Investigation Committee was established
	21-Aug	Airborne survey began
1959	20-Apr	Groundbreaking ceremony at the east exit of the Shin Tanna Tunnel
	10-Aug	Budget notification (1.86B $)
1960	5-May	World Bank Research Mission to Japan
1961	2-May	JNR President Sogo signed a World Bank loan ($80 M)
	4-Aug	Determination of construction standard outline
	15-Sep	Determination of prototype car specifications
	18-Oct	Determination of all routes
	1-Dec	Determination of rail weight and cross-sectional shape
1962	20-Jun	Completion of six prototype cars
	26-Jun	Test runs on the test section started (Sogo took a ride)
	30-Oct	World Bank Research Mission's test Ride
	30-Nov	Completion of the test Sect. (32 km)
1963	30-Mar	4-car Train B recorded 256 km/h
	20-May	JNR President Sogo retired. Chief engineer Shima retired on 30
	20-Jun	The decision was made to open 30 round trips a day with 12-car trains
1964	20-Jan	Completion of land acquisition for all routes
	8-Jul	The names of the trains were decided as "Hikari" and "Kodama"
	15-Aug	General Control Center was completed and ATC, CTC operations began
	1-Oct	The opening ceremony was held in the presence of Her Majesty the Emperor and Empress
		Started business with one Hikari (4-hour operation) and one Kodama (5 h) every hour
1965	1-Nov	Travel time reduced; Hikari to 3 h 10 min and Kodama to 4 h
	12-Oct	Receiving "Columbus Gold Award"

(continued)

© The Editor(s) (if applicable) and The Author(s), under exclusive license to Springer
Nature Singapore Pte Ltd. 2022
T. Shimomae, *Birth of the Shinkansen*,
https://doi.org/10.1007/978-981-16-6538-7

(continued)

Year	Day-Month	Event
1967	24-Mar	Receiving "Sperry Award"
	13-Jul	The number of passengers reached 100 million
	1-Oct	Timetable revision, 143 trains/day
1969	26-Jun	India's Prime Minister Indira Gandhi took a ride
	2-Aug	Former Chief Engineer Shima received James Watt Award
1972	15-Mar	Sanyo Shinkansen opened between Shin-Osaka and Okayama. The use of COMTRAC started
1975	10-Mar	Sanyo Shinkansen opened the whole route from Shin-Osaka to Hakata
	12-May	Queen Elizabeth of England took a ride
	29-Jul	Shinkansen environmental standards were announced
1976	25-May	The number of passengers reached one billion
	1-Jul	Timetable revision, Tokaido Shinkansen 275 trains/ day
1978	26-Oct	China's Vice Premier Deng Xiaoping took a ride
1980	30-May	China's Prime Minister Hua Guofeng took a ride
1981	27-Sep	(TGV opened at 260 km/h)
1982	23-Jun	Tohoku Shinkansen opened with Series 200 trains (Ueno-Morioka)
	15-Nov	Joetsu Shinkansen opened with Series 200 trains (Omiya-Niigata)
1984	3-Apr	The number of passengers of the Shinkansen reached two billion
1985	1-Feb	The fourth-generation COMTRAC started operation
	1-Oct	Series 100 trains started operation
1986	1-Nov	Timetable revision, Tokaido Shinkansen 310 trains/day
1987	1-Apr	JNR was privatized and split into JR companies
1989	11-Mar	Timetable revision, Tokaido Shinkansen 348 trains/day (JR Central)
1990	10-Mar	Timetable revision, Tokaido Shinkansen 373 trains/day (JR Central)
1991	16-Mar	Timetable revision, Tokaido Shinkansen 392 trains/day (JR Central)
	19-Apr	Soviet Union President Gorbachev took a ride
1992	14-Mar	Series 300 Nozomi started operating at 270 km/h (JR Central)
		The use of The Urgent Earthquake Detection and Alarm System (UrDAS) started
	1-Jul	Series 400 trains started operation between Tokyo, Fukushima, and Yamagata (JR East)
	8-Aug	Series 500 commercial vehicles recorded 350.4 km/h (JR West)
1993	21-Dec	STAR 21 test car reaches 425 km/h (JR East)
1994	15-Jul	Series E1 trains started operation on the Tohoku Shinkansen (JR East)
1995	17-Jan	"The Great Hanshin-Awaji Earthquake"
1996	26-Jul	300X test car recorded 443 km/h (JR Central)
	1-Oct	Joetsu Shinkansen started operating at 260 km/h (JR East)
1997	22-Mar	Sanyo Shinkansen started operating at 300 km/h with Series 500 cars (JR West)
		Series E2 and E3 coupled trains began operation between Tokyo and Morioka, of which E3 trains to Akita (JR East)

(continued)

(continued)

Year	Day-Month	Event
	1-Oct	Hokuriku Shinkansen opened between Tokyo and Nagano with Series E2 trains (JR East)
	20-Dec	Series E4 trains started operation on the Tohoku Shinkansen (JR East)
1999	13-Mar	Series 700 trains started operation on the Tokaido and Sanyo Shinkansen (JR Central and West)
2002	1-Dec	The use of the digitalized ATC began on the Tohoku Shinkansen (Morioka–Hachinohe) (JR East)
2004	13-Mar	Kyushu Shinkansen opened with Series 800 vehicles (Shin-Yatsushiro to Kagoshima-Chuo)
	23-Oct	"Mid Niigata Prefecture Earthquake in 2004"; A Shinkansen train derailed
2006	18-Mar	The use of the digitalized ATC began on the Tokaido Shinkansen (JR Central)
2007	1-Jul	Series N700 trains started operation on the Tokaido and Sanyo Shinkansen (JR Central and West)
2010	4-Dec	Tohoku Shinkansen opened the whole route from Tokyo to Shin-Aomori (JR East)
2011	5-Mar	Series E5 trains started operation at 300 km/h on the Tohoku Shinkansen (JR East)
	11-Mar	"The 2011 off the Pacific coast of Tohoku Earthquake"
2013	16-Mar	Series N700A trains started operation on the Tokaido and Sanyo Shinkansen (JR Central and West)
		Tohoku Shinkansen started 320 km/h operations with Series E5 cars (JR East)
		Series E6 trains started operation to Akita via Tohoku Shinkansen (JR East)
2014	15-Mar	Series E7 trains started operation on the Hokuriku Shinkansen (JR East)
2015	14-Mar	Tokaido Shinkansen started 285 km/h operation with Series N700A cars (JR Central)
		Hokuriku Shinkansen opened to Kanazawa (JR East and West)
2020	1-Jul	Series N700S trains started operation on the Tokaido and Sanyo Shinkansen (JR Central and West)

Index

© The Editor(s) (if applicable) and The Author(s), under exclusive license to Springer Nature Singapore Pte Ltd. 2022
T. Shimomae, *Birth of the Shinkansen*,
https://doi.org/10.1007/978-981-16-6538-7

Printed in the United States
by Baker & Taylor Publisher Services